THE METRIC SYSTEM
for
SECONDARY SCHOOLS

"Weights and measures may be ranked among the necessaries of life to every individual of human society. They enter into the economical arrangements and daily concerns of every family. They are necessary to every occupation of human industry; to the distribution and security of every species of property; to every transaction of trade and commerce; to the labors of the husbandman; to the ingenuity of the artificer; to the studies of the philosopher; to the researches of the antiquarian, to the navigation of the mariner, and the marches of the soldier; to all the exchanges of peace, and all the operations of war. The knowledge of them, as in established use, is among the first elements of education, and is often learned by those who learn nothing else, not even to read and write. This knowledge is riveted in the memory by the habitual application of it to the employments of men throughout life."

JOHN QUINCY ADAMS
Report to the Congress, 1821

THE METRIC SYSTEM
FOR
SECONDARY SCHOOLS

Compiled and Edited by

A. L. LeMaraic

and

J. P. Ciaramella

PUBLISHED BY ABBEY BOOKS

Metric Media Book Publishers

Somers, N.Y. 10589

ISBN Prefix 0-913768

THE METRIC SYSTEM FOR SECONDARY SCHOOLS

First Edition
L. C. Card Number 74-10176
ISBN Number 0-913768-04-9

Printed in the United States of America

FOREWORD

The need of a simplified version of the new metric system at the secondary school level actuated the construction of this book. Its purpose is to give expression to the fundamental and unchanging principles of the new metric system — Le Système International d'Unités, generally known as the SI.

Preoccupation with the technical and seemingly forbidding terminology of the metric system has too often served as a deterrent to many who are not aware of its simplicity, hence, the objective herein is to render a comprehensive presentation of the subject in a succinct and easy to understand format with practice exercises in well arranged sequence to lead on, step-by-step, to a better understanding of the new metric system and hasten its application in practical everyday use.

This book is based on the translation approved by the International Bureau of Weights and Measures of its publication "Le Système International d'Unités."

OTHER METRIC BOOKS

THE COMPLETE METRIC SYSTEM with THE INTERNATIONAL SYSTEM OF UNITS (SI)
L.C. Card Number 72-97799   ISBN Number 0-913768-00-6

THE TEACHER'S GUIDE TO THE METRIC SYSTEM
L.C. Card Number 74-3811   ISBN Number 0-913768-01-4

THE METRIC SYSTEM FOR BEGINNERS
L.C. Card Number 74-3812   ISBN Number 0-913768-02-2

THE METRIC ENCYCLOPEDIA
L.C. Card Number 74-9235   ISBN Number 0-913768-03-0

CHARTS and POSTERS FOR THE CLASSROOM

No. C9131—METRICS FOR THE CLASSROOM (Set of 5 Charts)
No. P9132—THE METRIC SYSTEM FOR EVERYDAY USE (1 Chart)
No. TP9133—METRIC CONVERSIONS AT A GLANCE (Set of 2 Posters) U.S. Customary to Metric — Metric to U.S. Customary
No. CP9134—COMBINATION SET (All the 8 Charts and Posters above)

TABLE OF CONTENTS

METHICS

Since 1952

INTRODUCTION

The value of everything we consume, use, trade, or exchange is determined by some means of measurement. We use measures in some manner every day of our lives in commerce, industry, science, education, and in our homes.

The rapid expansion of world trade, the enormous strides forward in modern science and technology, and the exchange of industrial and scientific techniques emphasized the need of a single worldwide, coordinated measurement system. As a result, in 1960, the General Conference on Weights and Measures adopted an extensive revision and simplification of the metric system for universal use after making official inquiry in scientific, technical, and educational circles in all countries seeking their opinions and recommendations regarding the establishment of a standardized system of measurement.

These efforts resulted in a modernized metric system approved by all countries. The new system was named "Le Système International d'Unités" — The International System of Units, with the international abbreviation SI, thus establishing a comprehensive specification for all units of measurement to be used universally.

This book is based upon the principles characterized in the new SI.

January 1975   A. L. LeMaraic
                                                J. P. Ciaramella

LIST OF TABLES

I. THE MODERNIZED METRIC SYSTEM

The International System of Units (SI) is the system of metric units adopted by the international metric authority, the General Conference on Weights and Measures (CGPM), and formalized by that body in 1960. The International Standards Organization (ISO) has adopted this new system for the expression of its standards and almost all countries of the world have accepted it as the sole system for trade and standardization. The SI is a coherent system of metric measurements based on the decimal system with distinctly defined basic units.

In its original conception the metre was the fundamental unit of the metric system. All units of length and capacity were to be derived directly from the metre which was intended to be equal to one-ten millionth of the earth's quadrant. It was also originally planned that the unit of mass, the kilogram, should be equal to the mass of a cubic decimetre of water at its maximum density. The units of length and mass are now defined independently of these original conceptions.

A new definition of the metre in terms of the wave length of light was adopted by the 11th General Conference on Weights and Measures in 1960. According to this new definition, 1 metre = 1 650 763.73 wave lengths in a vacuum of orange-red radiation of krypton 86.

The kilogram is independently defined as the mass of a cylinder of platinum-uranium alloy kept by the International Bureau of Weights and Measures at Paris. A duplicate in the custody of the National Bureau of Standards serves as the mass standard for the United States.

The litre is a secondary or derived unit of capacity or volume. It has been defined to be exactly equal to 1 cubic decimetre, i.e. 1 litre $=$ 1 dm³.

All lengths, area, and cubic measures which follow are derived from the international metre, the basic relation between units of the metric and customary system being:

1 metre $=$ 39.37 inches

From this relation it follows that 1 inch is exactly equal to 25.4 millimetres.

All capacities are based on the equivalent 1 litre equals 1 cubic decimetre (dm³). The decimetre is equal to 3.937 inches in accordance with the legal equivalent of the metre given above.

Measurements of mass are derived from the international kilogram as authorized. The relation used is 1 avoirdupois pound is equal to 453.592 grams.

In the construction of tables in this book, when the fundamental relation of the units furnished directly a reduction factor for determining the multiples of the units, this factor was used in its fundamental form, as, for example, 1 metre equals 39.37 inches. However, when the reduction factors had to be obtained by resorting to mathematical calculations the fundamental relations were carried out to a greater degree of accuracy than can ordinarily be measured thus making it possible to make computations to a higher degree of accuracy when desired. The basic relations between units of the metric and customary systems are given herein solely for academic purposes. Conversions should be avoided as much as possible with concentration on the use of metric measurements exclusively.

When the tables do not give the equivalent of any desired quantity directly and completely, it can usually be obtained without the necessity of making calculations by using quantities from several tables, shifting decimal points, if necessary, and merely adding the results.

Numerical prefixes are used with the words metre, gram, and litre to form supplementary units expressing quantities greater than a unit (multiples) or quantities smaller than a unit (submultiples). A distinction is made between the two by using Greek prefixes for multiples, and Latin prefixes for submultiples.

In the old British-American imperial system of measurement the names of units of length such as inch, foot, yard, furlong, mile, etc., do not indicate that they are units of length, and they certainly do not indicate the numerical relationship of one to the other.

In the SI, on the other hand, the name of any multiple or submultiple consists of

   1. the name of the principal unit which immediately

identifies it as a unit of length, mass, volume, capacity, force, etc., together with

2. a prefix, as in the table that follows. Each prefix has a fixed meaning which indicates the relationship between the multiple or submultiple and the principal unit. For instance, the name of the unit kilometre indicates, through the metre included in it, that it is a unit of length, and through the prefix kilo- that it is one thousand metres.

The names of multiples and submultiples of the units are formed by means of the following prefixes:

Prefix	Symbol	Multiply Unit by the Factor	
tera-	T	1 000 000 000 000	$= 10^{12}$
giga-	G	1 000 000 000	$= 10^{9}$
mega-	M	1 000 000	$= 10^{6}$
kilo-	k	1 000	$= 10^{3}$
(hecto-)	(h)	(100	$= 10^{2})$
(deka-)	(da)	(10	$= 10^{1})$
(deci-)	(d)	(0.1	$= 10^{-1})$
(centi-)	(c)	(0.01	$= 10^{-2})$
milli-	m	0.001	$= 10^{-3}$
micro-	μ	0.000 001	$= 10^{-6}$
nano-	n	0.000 000 001	$= 10^{-9}$
pico-	p	0.000 000 000 001	$= 10^{-12}$
femto-	f	0.000 000 000 000 001	$= 10^{-15}$
atto-	a	0.000 000 000 000 000 001	$= 10^{-18}$

HOW TO LEARN THE METRIC SYSTEM QUICKLY

The first step one must take to learn the metric system quickly is to start thinking in metric terms at once. Avoid making conversions from one system to the other as much as possible and make every effort to use only metric measurements.

The next step is to familiarize oneself with the new SI units of measure most commonly used and their multiples, submultiples, and symbols. The units most commonly used are the metre (m), the litre (l), and the kilogram (kg). Get acquainted with the quantities they represent by observing the items marked in metric measurements which are appearing more and more on the shelves at supermarkets and other establishments. You may be surprised at the large number of things already carrying metric measurements which you may have overlooked. Making these observations will acquaint you with the quantities you will need when buying by metric measurement in the future.

Compare and learn the relationship between the new metric measures and the old ones you have been accustomed to use. Provide yourself with metric scales and measuring utensils, metric measuring-sticks, rulers, tape-measures, and, last but not least, a bathroom scale in metric.

Use the metric system at every opportunity to measure things you buy, such as drapes, carpets, sheets, pillow cases, bedspreads, fabrics, clothing, and also start to think of your fuel oil, gasoline, road speeds and distances in metric terms. The metric system is here to stay; it is easy to use, and if you start to think metric at once you will be surprised how quickly you will master it.

THE METRIC SYSTEM FOR EVERYDAY USE

1. LEARN ONLY TWELVE TERMS

By learning the meaning of 12 terms, viz. mega-, kilo-, deci-, centi-, milli-, metre, hectare, litre, gram, metric ton, pascal and degree Celsius, enough knowledge of the metric system is acquired to satisfy practically all ordinary needs. Please note that according to international agreement there is only one correct symbol for each term. These symbols are the same for all languages and are given in brackets below. They have no plurals and are consistently written without a period. The units of electricity and time are not affected by metrication.

2. METRIC PREFIXES

mega-	(M)	means	million
kilo-	(k)	"	thousand
deci-	(d)	"	tenth
centi-	(c)	"	hundredth
milli-	(m)	"	thousandth

These prefixes are the same for all types of metric units,

1 megalitre (Ml)	=	1 000 000 litres (l)
1 kilolitre (kl)	=	1 000 litres (l)
1 kilopascal (kPa)	=	1 000 pascals (Pa)
1 kilogram (kg)	=	1 000 grams (g)
1 kilometre (km)	=	1 000 metres (m)
1 decimetre (dm)	=	1/10th metre (m)
1 centimetre (cm)	=	1/100th metre (m)
1 millimetre (mm)	=	1/1 000th metre (m)

3. THE METRIC SYSTEM FOR EVERYDAY USE
(continued)

(a) LENGTH, DISTANCE

Base unit:
metre (m)   approximately equal to a long pace

Multiple:
kilometre (km)   = 1 000 metres (m)

Submultiples:
centimetre (cm)   = 1/100th metre (m)
millimetre (mm)   = 1/1 000th metre (m)

(b) AREA

SI unit:
square metre (m²)   = the area of a square with
                                         sides 1 metre

Multiples:
1 square kilometre (km²)   = 1 000 000 square metres
                                            (m²)
1 hectare (ha)   = 10 000 square metres (m²)

Submultiples:
1 square centimetre
    (cm²)   = 1/10 000th square metre
                                            (m²)
1 square millimetre (mm²) = 1/1 000 000th square
                                            metre (m²)

Note: Hectare (ha) is a special name given to 10 000 m² and is
           used in land surveying.

THE METRIC SYSTEM FOR EVERYDAY USE
(continued)

(c) VOLUME

    SI unit:
        cubic metre (m^3)   = the volume of a cube
                                          with sides 1 metre

    Submultiples:
        1 cubic decimetre (dm^3)   = 1/1 000th cubic metre
                                            (m^3)
        1 cubic centimetre (cm^3)   = 1/1 000 000th cubic
                                            metre (m^3)

Note: The cubic decimetre (dm^3) was given a special name, the litre (l). The litre is used for measuring liquids.

    Unit for liquids:
        1 litre (l)   = 1 cubic decimetre (dm^3)

    Multiples:
        1 megalitre (Ml)   = 1 000 000 litres (l)
                                 = 1 000 cubic metres (m^3)
        1 kilolitre (kl)   = 1 000 litres (l)
                                 = 1 cubic metre (m^3)

    Submultiple:
        1 millilitre (ml)   = 1/1 000 litre (l)
                                 = 1 cubic centimetre (cm^3)

(d) MASS

    Base unit:
        kilogram (kg)   = the mass of 1 litre (l) of water

    Multiple:
        1 metric ton (t)   = 1 000 kilograms (kg)

    Submultiples:
        1 gram (g)   = 1/1 000th kilogram (kg)
        1 milligram (mg)   = 1/1 000th gram (g)
NOTE: The megagram (Mg) = 1 000 kilograms (kg) and is
        called a metric ton (t).

THE METRIC SYSTEM FOR EVERYDAY USE
(continued)

(e) VELOCITY

Expressed in kilometres per hour (km/h) (h is derived from the Latin word "hora").

(f) PRESSURE

For practical purposes pressure is expressed in kilopascal (kPa). 100 kPa is practically equal to the pressure of the atmosphere at sea level.

SI unit:
   pascal (Pa)   = a pressure of 1 newton per square metre (1 N/m²)

Multiples:
   1 megapascal (MPa) = 1 000 000 pascals (Pa)
   1 kilopascal (kPa) = 1 000 pascals (Pa)

(g) TEMPERATURE

Temperature is expressed in degrees Celsius (⁰C). Water freezes at 0 ⁰C and boils at 100 ⁰C. Body temperature is 37 ⁰C and 20 ⁰C is a pleasant room temperature.

(h) FUEL CONSUMPTION

Expressed in litres per 100 kilometres (1/100 km)

METRIC TABLES, SYMBOLS AND NOTATION

1. LENGTH

10 millimetres (mm)	=	1 centimetre (cm)
100 cm	=	1 metre (m)
1 000 m	=	1 kilometre (km)

2. AREA

100 mm²	=	1 cm²
100 cm²	=	1 dm² (square decimetre)
100 dm²	=	1 m²
100 m²	=	1 are (a)
100 ares	=	1 hectare (ha)
100 ha	=	1 km²

A piece of land 100 metres by 100 metres is therefore 1 hectare.

3. VOLUME

1 000 millilitres (ml)	=	1 000 cm³	=	1 litre (l)
1 000 litres (l)	=	1 m³		

4. MASS

1 000 grams (g)	=	1 kilogram (kg)
1 000 kg	=	1 metric ton (t)

METRIC TABLES, SYMBOLS AND NOTATIONS
(continued)

5. PREFERRED UNITS, MULTIPLES AND SUBMULTIPLES

Length: 1 000 millimetres (mm) = 1 metre (m);
1 000 m = 1 kilometre (km)
Volume: 1 000 millilitres (ml) = 1 litre (l);
1 000 litres = 1 m³ = 1 kilolitre (kl)
Mass: 1 000 grams (g) = 1 kilogram (kg);
1 000 kg = 1 metric ton (t)

Writing numbers: Large numbers should be divided into groups of three, counting from the decimal sign to the left and the right and these groups should be separated by a space and never by a comma. A space replaces the comma used previously because the comma serves as the decimal sign in many metric countries. In numbers less than units a zero should precede the decimal sign, for example, 0.239 and not .239.

A quantity is generally described in millimetres, grams or millilitres up to 999.9, thereafter in metres, kilograms or litres. 1 300 mm or g or ml is therefore written as 1.3 m (1.300 m for engineering drawings), 1.3 kg or 1.3 litres. When using symbols, no addition such as s, should be made to indicate plurality, e.g. 1 kg; 2 kg NOT 2 kgs.

It is important to use the symbols correctly, lower case letters should not be used where the accepted symbol is a capital letter, or vice versa, e.g. m stands for milli- (1/1 000 th), but M for mega- (1 000 000 times).

II. DEFINITIONS OF UNITS

1. Length

Fundamental Units

A metre (m) is a unit of length equal to 1 650 763.73 wave lengths in a vacuum of orange-red radiation of krypton 86.

Multiples and Submultiples

1 kilometre (km)	=	1 000 metres.
1 hectometre (hm)	=	100 metres.
1 dekametre (dam)	=	10 metres.
1 decimetre (dm)	=	0.1 metre.
1 centimetre (cm)	=	0.01 metre.
1 millimetre (mm)	=	0.001 metre.
1 micrometre (μm)	=	0.000 001 metre
	=	0.001 millimetre.
1 nanometre (nm)	=	0.000 000 001 metre
	=	0.001 micrometre.
	=	0.000 000 1 millimetre.
1 angstrom (A)	=	0.000 1 micrometre.
	=	0.1 nanometre

DEFINITIONS OF UNITS (continued)

2. Area

Fundamental Units

A square metre (m^2) is a unit of area equal to the area of a square the sides of which are 1 metre.

Multiples and Submultiples

1 square kilometre (km^2)	= 1 000 000 square metres.
1 hectare (ha), or square hectometre (hm^2)	= 10 000 square metres.
1 are (a), or square dekametre (dam^2)	= 100 square metres.
1 centare (ca)	= 1 square metre.
1 square decimetre (dm^2)	= 0.01 square metre.
1 square centimetre (cm^2)	= 0.000 1 square metre.
1 square millimetre (mm^2)	= 0.000 001 square metre.

DEFINITIONS OF UNITS (continued)

3. Volume

Fundamental Units

A cubic metre (m^3) is a unit of volume equal to a cube the edges of which are 1 metre.

Multiples and Submultiples

1 cubic kilometre (km^3)	= 1 000 000 000 cubic metres.
1 cubic hectometre (hm^3)	= 1 000 000 cubic metres.
1 cubic dekametre (dam^3)	= 1 000 cubic metres.
1 cubic decimetre (dm^3)	= 0.001 cubic metre.
1 cubic centimetre (cm^3)	= 0.000 001 cubic metre
	= 0.001 cubic decimetre.
1 cubic millimetre (mm^3)	= 0.000 000 001 cubic metre
	= 0.001 cubic centimetre.

4. Capacity

Fundamental Units

A litre (l) is a unit of capacity equal to 1 cubic decimetre (dm^3).

Multiples and Submultiples

1 hectolitre (hl)	= 100 litres.
1 dekalitre (dal)	= 10 litres.
1 decilitre (dl)	= 0.1 litre.
1 centilitre (cl)	= 0.01 litre.
1 millilitre (ml)	= 0.001 litre
	= 1 cubic centimetre (cm^3)

DEFINITIONS OF UNITS (continued)

5. Mass

Fundamental Units

A kilogram (kg) is a unit of mass equal to the mass of the International Prototype Kilogram.
A gram (g) is a unit of mass equal to one-thousandth of the mass of the International Prototype Kilogram.

Multiples and Submultiples

1 metric ton (t)	=	1 000 kilograms.
1 hectogram (hg)	=	100 grams.
1 dekagram (dag)	=	10 grams.
1 decigram (dg)	=	0.1 gram.
1 centigram (cg)	=	0.01 gram.
1 milligram (mg)	=	0.001 gram.

III. LENGTH

Length is a measure of extent or distance. The metre (m) with its recommended multiples and submultiples, the kilometre (km), and millimetre (mm) will be used as units of length in trade, industry, science, and other fields. However, there are some exceptions where owing to practical considerations the centimetre (cm) will be used, such as, textiles, clothing, body measurements, and many other things.

Units

The units of length which are of primary importance are:

SI principal unit	Recommended SI multiples and submultiples	Other SI multiples and submultiples
metre (m)	kilometre (km) 1 km = 1 000 m	centimetre (cm) $1 \text{ cm} = \frac{1}{100} \text{ m}$
	millimetre (mm) $1 \text{ mm} = \frac{1}{1\,000} \text{ m}$ $= 0.001 \text{ m}$	

EQUIVALENTS

Measure of Length

10 millimetres (mm)	= 1 centimetre (cm)
10 centimetres (cm)	= 1 decimetre (dm)
10 decimetres (dm)	= 1 metre (m)
10 metres (m)	= 1 dekametre (dam)
10 dekametres (dam)	= 1 hectometre (hm)
10 hectometres (hm)	= 1 kilometre (km)

1 centimetre (cm)	= 10 millimetres (mm)
1 metre (m)	= 100 centimetres (cm)
	= 1 000 millimetres (mm)
1 kilometre (km)	= 1 000 metres (m)

The above table shows that each unit of length is 10 times larger than its next smaller unit, and inversely, it is 1/10 of its next larger unit.

We can change from one unit to another by multiplying or dividing by 10, 100, 1 000 or other power of 10 according to the conversion desired, up or down.

Rule 1.
To change from any given unit to any smaller unit **multiply** the given unit by the proper conversion factor or factors shown in the table above.

Example:

Rule: Given unit x conversion factor = smaller unit
Problem: Change 5 metres to decimetres.
Method: The given unit is metres.
           The conversion factor is 10 because 10 dm = 1 m
thus,
           5 m × 10 = 50 dm

Rule 2.
To change from any given unit to any larger unit **divide** the given unit by the proper conversion factor or factors shown in the table above.

Example:

Rule:
Given unit ÷ conversion factor = larger unit.

Problem:
Change 50 decimetres to metres.

Method:
The given unit is decimetres.

The conversion factor is 10 because 10 dm = 1 m thus,

50 dm ÷ 10 = 5 m

PRACTICE EXAMPLES - MEASURE OF LENGTH

Exercise 1

Convert to millimetres:	Convert to centimetres:	Convert to decimetres:	Convert to centimetres:
Group 1	Group 2	Group 3	Group 4
a. 95 cm	a. 8.25 dm	a. 30 m	a. 8 m
b. 23.5 cm	b. 40 dm	b. 0.625 m	b. 0.75 m
c. 8 cm	c. 6.20 dm	c. 8 m	c. 7.60 m
d. 9.5 cm	d. 9 dm	d. 18.45 m	d. 15 m
e. 12 cm	e. 3.25 dm	e. 3 m	e. 65 m

Exercise 2

Convert to millimetres:	Convert to metres:	Convert to dekametres:	Convert to hectometres:
Group 5	Group 6	Group 7	Group 8
a. 9 m	a. 20 dam	a. 10 hm	a. 5 km
b. 15.5 m	b. 4 dam	b. 9.25 hm	b. 16 km
c. 25 m	c. 75 dam	c. 45 hm	c. 63 km
d. 6.33 m	d. 8.4 dam	d. 2.625 hm	d. 0.25 km
e. 12.125 m	e. 4.64 dam	e. 17 hm	e. 8.63 km

PRACTICE EXAMPLES - MEASURE OF LENGTH
(continued)

Exercise 3

Convert to metres:	Convert to metres:	Convert to centimetres:	Convert to decimetres:
Group 9	Group 10	Group 11	Group 12
a. 6 hm	a. 5 km	a. 45 mm	a. 30 cm
b. 15.2 hm	b. 53 km	b. 21.7 mm	b. 55 cm
c. 24 hm	c. 121 km	c. 16 mm	c. 8 cm
d. 0.375 hm	d. 37.5 km	d. 875 mm	d. 25.3 cm
e. 89 hm	e. 0.125 km	e. 32.5 mm	e. 0.6 cm

Exercise 4

Convert to metres:	Convert to metres:	Convert to metres:	Convert to dekametres:
Group 13	Group 14	Group 15	Group 16
a. 125 dm	a. 350 cm	a. 2 500 mm	a. 75 m
b. 15 dm	b. 222 cm	b. 1 250 mm	b. 185.5 m
c. 4 dm	c. 1 624 cm	c. 10 375 mm	c. 29.5 m
d. 7.25 dm	d. 635.5 cm	d. 239.5 mm	d. 8 m
e. 12.5 dm	e. 33 cm	e. 55.5 mm	e. 6.3 m

PRACTICE EXAMPLES - MEASURE OF LENGTH
(continued)

Exercise 5

Convert to hectometres:	Convert to kilometres:	Convert to hectometres:	Convert to kilometres:
Group 17	Group 18	Group 19	Group 20
a. 175 dam	a. 65 hm	a. 550 m	a. 350 m
b. 38 dam	b. 7 hm	b. 148.7 m	b. 7 850 m
c. 6 dam	c. 16.25 hm	c. 340 m	c. 309.2 m
d. 57.5 dam	d. 225.5 hm	d. 2 125 m	d. 1 728 m
e. 0.9 dam	e. 29 hm	e. 44 m	e. 2 000 m

Exercise 6

Convert to centimetres:	Convert to kilometres:	Convert to millimetres:	Convert to kilometres:
Group 21	Group 22	Group 23	Group 24
a. 3 km	a. 650 000 cm	a. 15 km	a. 3 500 000 mm
b. 0.625 km	b. 125 000 cm	b. 4 km	b. 25 000 mm
c. 18 km	c. 253 500 cm	c. 2.25 km	c. 750 000 mm
d. 7.25 km	d. 45 000 cm	d. 0.5 km	d. 900 000 mm
e. 1.625 km	e. 178 400 cm	e. 1.75 km	e. 400 000 mm

PRACTICE EXAMPLES - MEASURE OF LENGTH
(continued)

Exercise 7

Give the following answers:

Group 25

    a. 5 km 8 m   =   m ?
    b. 3 m 7 cm   =   cm ?
    c. 4 cm 5 mm   =   mm ?
    d. 1 m 1 cm 1 mm   =   mm?
    e. 5 m 5 dm 5 cm   =   cm ?

Group 26

    a. 6 m 42 cm   =   cm?
    b. 5 m 50 cm   =   m ?
    c. 6 km 250 mm   =   km ?
    d. 3 m 2 dm 6 cm 3 mm   =   m ?
    e. 25 cm 10 mm   =   cm?

Add the following:

Group 27

    a. 7 km + 2 hm + 3 dam +
            4 dm + 6 cm + 5 mm   =   m ?
    b. 15.25 km + 5 hm + 4 dam + 11 m   =   m ?
    c. 6m + 6 dm + 6 cm + 6 mm   =   m ?
    d. 3 km + 2 hm + 1 dam + 6 m   =   m ?
    e. 2 km + 5 hm + 1 dam +
            3 m + 8 dm + 4 cm + 7 mm   =   m ?

Group 28

    a. 5 km + 2 dm + 12 dm + 5 mm   =   m ?
    b. 7 hm + 3 m + 10 cm   =   m ?
    c. 1 km + 1 hm + 1 dam +
            1 m + 1 dm + 1 cm + 1 mm   =   m?
    d. 1 m + 5 dm + 50 cm + 250 mm   =   m ?
    e. 2.5 km + 1.25 hm + 2.5 dam + 6.5 m   =   m ?

PRACTICE EXAMPLES - MEASURE OF LENGTH
(continued)

Exercise 8

Give the following measurements in metres:

Group 29

a. 6 km 2 hm 1 dam 7 m 4 dm 2 cm 3 mm   =   m ?

b. 14 km 5 hm 2 dam 6 m   =   m ?

c. 11 m 7 dm 4 cm 8 mm   =   m ?

d. 6 dam 6 dm   =   m ?

e. 2 km 5 hm 1 dam 3 m 5 dm 8 cm 5 mm   =   m ?

Give the following measurements in metres:

Group 30

a. 4 km 4 hm 4 dam 4 m   =   m ?

b. 5 m 8 dm 3 cm 4 mm   =   m ?

c. 5 hm 5 dam 5 dm 5 cm 5 mm   =   m ?

d. 15 km 2 hm 1 dam 2 m 5 dm 3 cm 2 mm   =   m ?

e. 5 km 4 hm 3 dam 2 m and 5 mm   =   m ?

PRACTICE EXAMPLES - MEASURE OF LENGTH
(continued)

Exercise 9

Convert:

Group 31

a. 16 cm to mm
b. 2.6 dm to cm
c. 92 m to dm
d. 3.5 m to cm
e. 3.2 m to mm

Convert:

Group 32

a. 4 dam to m
b. 15.5 hm to dam
c. 31 km to hm
d. 3.75 hm to m
e. 9 km to m

Convert:

Group 33

a. 25 mm to cm
b. 40 cm to dm
c. 11 dm to m
d. 600 cm to m
e. 7 500 mm to m

Convert:

Group 34

a. 7.5 m to dam
b. 17 dam to hm
c. 7.5 hm to km
d. 112 m to hm
e. 3 255 m to km

IV. AREA

Area is the surface contents of any figure determined by measuring the number of units of square measure contained within its perimeter. The area of a rectangle is equal to its length times its width.

When computing area the linear units of measure used should be in the same denomination, i.e. millimetres $\times$ millimetres; centimetres $\times$ centimetres; decimetres $\times$ decimetres, etc. There are various formulae for computing areas of geometric figures. These formulae remain the same when used with metric measurements.

Units

The units of area which are of primary importance are:

SI principal unit	SI multiples and submultiples	Decimal multiples of SI units with special names
	square kilometre (km²) $1 \text{ km}^2 = 1\ 000\ 000 \text{ m}^2$	
		hectare (ha) $1 \text{ ha} = 10\ 000 \text{ m}^2$
square metre (m²)		
	square centimetre (cm²) $1 \text{ cm}^2 = 0.000\ 1 \text{ m}^2$ $= 1/10\ 000 \text{ m}^2$ square millimetre (mm²) $1 \text{ mm}^2 = 0\ 000\ 001 \text{ m}^2$ $= 1/1\ 000\ 000 \text{ m}^2$	

Measure of Area

100 square millimetres (mm²)	=	1 square centimetre (cm²)
100 square centimetres	=	1 square decimetre (dm²)
100 square decimetres	=	1 square metre (m²)
100 square metres	=	1 square dekametre (dam²)
		or 1 are (a)
100 square dekametres	=	1 square hectometre (hm²)
		or 1 hectare (ha)
		= 10 000 m²
100 square hectometres	=	1 square kilometre (km²)
(or 100 hectares)		= 1 000 000 m²)

It is obvious from the table above that each unit of area is 100 times larger than its next smaller unit, and inversely, it is 1/100 of its next larger unit. It follows therefore, that we can change from one unit of area to another by multiplying or dividing by 100, 1 000, 10 000 or other power, according to the conversion desired, up or down, as stated in Rules 1 and 2 on pages 18 and 19.

PRACTICE EXAMPLES - MEASURE OF AREA

Exercise 10

Convert:

Group 35

a. 5 cm² to mm²
b. 65 dm² to cm²
c. 20 m² to dm²
d. 4.6 m² to cm²
e. 16 m² to mm²

Convert:

Group 36

a. 150 mm² to cm²
b. 698 cm² to dm²
c. 5 dm² to m²
d. 55 000 cm² to m²
e. 523 000 mm² to m²

Convert:

Group 37

a. 0.75 km² to m²
b. 3 km² to m²
c. 5.5 km² to m²
d. 1.3 km² to m²
e. 1 km² to m²

Convert:

Group 38

a. 175 000 m² to km²
b. 250 000 m² to km²
c. 125 000 m² to km²
d. 62 500 m² to km²
e. 31 250 m² to km²

PRACTICE EXAMPLES - MEASURE OF AREA
(continued)

Exercise 11

Convert:

Group 39

a. 11 cm² to mm²
b. 1.75 dm² to cm²
c. 35 m² to dm²
d. 8 m² to cm²
e. 0.75 m² to mm²

Convert:

Group 40

a. 372 mm² to cm²
b. 88 cm² to dm²
c. 158.75 dm² to m²
d. 8 400 cm² to m²
e. 600 000 mm² to m²

Convert:

Group 41

a. 4.25 km² to m²
b. 5.375 km² to m²
c. 0.625 km² to m²
d. 1.75 km² to m²
e. 3.375 km² to m²

Convert:

Group 42

a. 2 250 000 m² to km²
b. 1 300 000 m² to km²
c. 100 000 m² to km²
d. 37 500 m² to km²
e. 50 000 m² to km²

V. VOLUME

The concept of volume is the space occupied by a quantity of matter as measured by cubic units. The cubic metre (m^3) is a unit of volume equal to a cube the edges of which are 1 metre. The numerical quantity values between units of volume derived from the millimetre, the metre, and the kilometre are very large, for instance,

$$1\ m^3 = 1\ 000\ mm \times 1\ 000\ mm \times 1\ 000\ mm = 1\ 000\ 000\ 000\ mm^3.$$

As a result, other submultiples, such as the cubic centimetre (cm^3) have to be used in order to avoid unmanageably large numbers.

The litre was defined as being precisely equal to one cubic decimetre (dm^3) (1 000 cubic centimetres).

The correct symbol for cubic centimetres is cm^3 and not cc or c.c. In the SI cc would stand for centi-centi, which is meaningless.

Units

The units of volume which are of primary importance are:

SI principal unit	SI multiples and submultiples	Decimal multiples of SI units with special names
cubic metre (m^3)	cubic centimetre (cm^3) $1\ cm^3 = 0\ 000\ 001\ m^3$ $= \dfrac{1}{1\ 000\ 000}\ m^3$ cubic millimetre (mm^3) $1\ mm^3 = 0.000\ 000\ 001\ m^3$ $= \dfrac{1}{1\ 000\ 000\ 000}\ m^3$	kilolitre (kl) $1\ kl = 1\ 000$ litres $= 1\ m^3$ litre (l) 1 litre $= 1\ 000\ cm^3$ $(\ =\ 1\ dm^3)$ millilitre (ml) $1\ ml = 0,001$ litres $= 1\ cm^3$

Measure of Volume

1 000 cubic millimetres (mm³)	=	1 cubic centimetre (cm³)
1 000 cubic centimetres (cm³)	=	1 cubic decimetre (dm³)
1 000 cubic decimetres (dm³)	=	1 cubic metre (m³)
1 000 cubic metres (m³)	=	1 cubic dekametre (dam³)
1 000 cubic dekametres (dam³)	=	1 cubic hectometre (hm³)
1 000 cubic hectometres (hm³)	=	1 cubic kilometre (km³)

It is obvious from the above table that each unit of volume (cubic measure) is 1 000 times larger than its next smaller unit, and inversely, it is 1/1 000 of its next larger unit. It follows therefore, that we can change from one unit of volume to another by multiplying or dividing by 1 000, 10 000, 100 000 or other power, according to the conversion desired, up or down. (See Rules 1 and 2 on pages 18 and 19.

Metric units of volume differ by 1 000 for this reason: 1 centimetre (cm) = 10 millimetres (mm). When these are cubed by multiplying length by width by height to get the measure of volume or cubic space occupied, we get 1 cm × 1 cm × 1 cm equals 1 cubic centimetre (cm³) which is the same as 10 mm × 10 mm × 10 mm equal to 1 000 cubic millimetres (mm³).

PRACTICE EXAMPLES - MEASURE OF VOLUME

Exercise 12

Convert:
Group 43

a. 4 cm³ to mm³
b. 17 dm³ to cm³
c. 12.5 m³ to dm³
d. 18 m³ to cm³
e. 7 m³ to mm³

Convert:
Group 44

a. 40 cm³ to mm³
b. 0.6 dm³ to cm³
c. 50 m³ to dm³
d. 5.5 m³ to cm³
e. 1.375 m³ to mm³

Convert:
Group 45

a. 2 000 mm³ to cm³
b. 650 cm³ to dm³
c. 2 520 dm³ to m³
d. 7 000 000 cm³ to m³
e. 6 500 000 mm³ to m³

Convert:
Group 46

a. 3 250 mm³ to cm³
b. 37.5 cm³ to dm³
c. 8 000 dm³ to m³
d. 35 000 cm³ to m³
e. 7 000 000 mm³ to m³

VI. CAPACITY

The litre (l) is a unit of liquid capacity equal in volume to exactly 1 cubic decimetre (dm^3).

Measure of Capacity

10 millilitres (ml)	= 1 centilitre (cl)
10 centilitres (cl)	= 1 decilitre (dl)
10 decilitres (dl)	= 1 litre
10 litres	= 1 dekalitre (dal)
10 dekalitres (dal)	= 1 hectolitre (hl)
10 hectolitres (hl)	= 1 kilolitre (kl)

It is obvious from the above table that each unit of capacity is 10 times larger than its next lower unit, and inversely, it is 1/10 of its next larger unit. It follows therefore, that we can change from one unit to another by multiplying or dividing by 10, 100, 1 000 or any other power of 10 according to the conversion desired, up or down, as stated in Rules 1 and 2 on pages 18 and 19.

PRACTICE EXAMPLES - MEASURE OF CAPACITY

Exercise 13

Convert:

Group 47

a. 20 dal to l
b. 25 hl to dal
c. 6.4 kl to hl
d. 37 hl to l
e. 42 kl to l

Convert:

Group 48

a. 18 ml to cl
b. 50 cl to dl
c. 7 dl to l
d. 600 cl to l
e. 5 875 ml to l

Convert:

Group 49

a. 5 cl to ml
b. 12 dl to cl
c. 15 l to dl
d. 48 l to cl
e. 6.6 l to ml

Convert:

Group 50

a. 112 l to dal
b. 2.77 dal to hl
c. 45 hl to kl
d. 250 l to hl
e. 86 l to kl

PRACTICE EXAMPLES - MEASURE OF CAPACITY
(continued)

Exercise 14

Convert:

Group 51

a. 20 cl to ml
b. 5 dl to cl
c. 0.75 l to dl
d. 15 l to cl
e. 25 l to ml

Convert:

Group 52

a. 9 ml to cl
b. 70 cl to dl
c. 21.3 dl to l
d. 75 cl to l
e. 2 500 ml to l

Convert:

Group 53

a. 7 dal to l
b. 7.5 hl to dal
c. 30 kl to hl
d. 15.25 hl to l
e. 50 kl to l

Convert:

Group 54

a. 7 l to dal
b. 56 dal to hl
c. 24.5 hl to kl
d. 45 l to hl
e. 4 950 l to kl

VII. MASS, FORCE, AND WEIGHT

The term weight has been used in many ways, resulting in confusion. In our everyday lives it is used as being synonymous with mass. In this usage it is measured in kilograms or grams. In physics and technology, however, it has been used as a force related to gravity. In that usage, the weight of an object stationary on the earth was defined as being equal to the force of gravity on the object and was measured in newtons. That use of the term weight is now falling into disfavor.

In the study of physics you will learn more about force. For now remember that it is measured in newtons and **not** in kilograms. Because weight was in the past often used as a force, a brief explanation of that usage will be of interest to you. The weight as a force of a body can be visualized as being the force of reaction that the body exerts on all forces that are in contact with it. This type of visualization will better enable you to see why it is that an astronaut when he is in space "falling" toward the earth or during his long voyage in skylab is said to be "weight-less"; i.e., that his weight is zero. This obviously does not mean that his mass is zero. It means only that except for gravitation there are no forces acting on him against which to react. He thus feels "weightless" even though he is not "massless."

Because of this confusion, the term weight will more and more be avoided in science and technology.

MASS

Mass is related to heaviness. The principal unit, the kilogram (kg), already uses a prefix in contrast to all other principal units. This is specifically mentioned to emphasize the fact that, in spite of the prefix, the kilogram is the principal unit of more common usage than the gram.

The term "megagram" will not be used very much as in most cases the term "metric ton" will be used instead. The old imperial ton we have used heretofore will be gradually supplanted by metric ton.

Units

The important units of mass are the following:

SI principal unit	Recommended SI multiples and submultiples	Other units
kilogram (kg)	megagram (Mg) 1 Mg = 1 000 kg gram (g) $1\,g = \dfrac{1}{1\,000}\,(kg)$ $= 0.001\ kg$ milligram (mg) $1\,mg = \dfrac{1}{1\,000\,000}\ kg$ $= 0.000\,001\ kg$	metric ton(t) 1 t = 1 000 kg

Measure of Mass

10 milligrams (mg)	=	1 centigram (cg)
10 centigrams (cg)	=	1 decigram (dg)
10 decigrams (dg)	=	1 gram (g)
10 grams (g)	=	1 dekagram (dag)
10 dekagrams (dag)	=	1 hectogram (hg)
10 hectograms (hg)	=	1 kilogram (kg)
1 000 kilograms (kg)	=	1 metric ton (t)

It is obvious from the above table that each unit of mass is 10 times larger than its next smaller unit, and inversely, it is 1/10 of its next larger unit. It follows therefore, that we can change from one unit of mass to another by multiplying or dividing by 10, 100, 1 000 or other power of 10, according to the conversion desired, up or down, as stated in Rules 1 and 2 on pages 18 and 19.

PRACTICE EXAMPLES - MEASURE OF MASS

Exercise 15

Convert:

Group 55

a. 5 cg to mg
b. 23 dg to cg
c. 12 g to dg
d. 60 g to cg
e. 0.625 g to mg

Convert:

Group 56

a. 40 mg to cg
b. 15 cg to dg
c. 7 dg to g
d. 150 cg to g
e. 2 000 mg to g

Convert:

Group 57

a. 8.75 hg to dag
b. 36 kg to hg
c. 15 hg to g
d. 3.75 kg to g
e. 4 t to kg

Convert:

Group 58

a. 125 dag to hg
b. 25.25 hg to kg
c. 2 500 g to hg
d. 3 200 g to kg
e. 22 250 kg to t

PRACTICE EXAMPLES - MEASURE OF MASS
(continued)

Exercise 16

Convert:
Group 59

a. 5.25 cg to mg
b. 25 dg to cg
c. 8 g to dg
d. 17 g to cg
e. 6.5 g to mg

Convert:
Group 60

a. 4 mg to cg
b. 45 cg to dg
c. 7.25 dg to g
d. 1 500 cg to g
e. 85 mg to g

Convert:
Group 61

a. 35 hg to dag
b. 25 kg to hg
c. 12.25 hg to g
d. 9 kg to g
e. 25 t to kg

Convert:
Group 62

a. 25.375 dag to hg
b. 25 hg to kg
c. 555 g to hg
d. 3 250 g to kg
e. 6 500 kg to t

VIII. VOLUME, CAPACITY, MASS

The relationship between some units of volume, capacity, and mass is given hereunder.

> 1 cubic centimetre (cm^3) of water has a mass equivalent of 1 gram (g).
> 1 millilitre (ml) of water has a mass equivalent of 1 gram (g).
> 1 cubic decimetre (dm^3) of water has a mass equivalent of 1 kilogram (kg).
> 1 litre (l) of water has a mass equivalent of 1 kilogram (kg).

MISCELLANEOUS EQUIVALENTS

1 litre = 1 cubic decimetre (dm^3) = 1 000 cubic centimetres (cm^3)
1 millilitre (ml) = 1 cubic centimetre (cm^3)
1 litre of water weighs 1 kilogram (kg)
1 millilitre (ml) or cubic centimetre (cm^3) of water weighs 1 gram (g)

PRACTICE EXAMPLES -
VOLUME, CAPACITY, AND MASS

Exercise 17

Give the equivalent units of volume:

Group 63	Group 64
a. 5 ml	a. 8 cl
b. 16 ml	b. 19 cl
c. 35 ml	c. 4 cl
d. 49 ml	d. 0.75 cl
e. 7.1 ml	e. 12.5 cl

Give the equivalent units of volume:

Group 65	Group 66
a. 4 dl	a. 2 l
b. 22 dl	b. 75 l
c. 80 dl	c. 40 l
d. 7.5 dl	d. 2.8 l
e. 42.25 dl	e. 12.75 l

PRACTICE EXAMPLES - VOLUME, CAPACITY, AND MASS (continued)

Exercise 18

Give the equivalent units of capacity:

Group 67	Group 68
a. 4 cm³	a. 25 mm³
b. 19 cm³	b. 15 mm³
c. 40 cm³	c. 4 mm³
d. 6.5 cm³	d. 2.5 mm³
e. 37.55 cm³	e. 7.75 mm³

Give the equivalent units of capacity

Group 69	Group 70
a. 4 dm³	a. 5 m³
b. 25 dm³	b. 7 m³
c. 46 dm³	c. 19 m³
d. 5.75 dm³	d. 0.875 m³
e. 20.375 dm³	e. 13.5 m³

PRACTICE EXAMPLES - VOLUME, CAPACITY, AND MASS (continued)

Exercise 19

Give the mass quantity of the following volumes of water:

Convert to kilograms: Group 71	Convert to grams: Group 72
a. 5 dm³	a. 7 dm³
b. 22 dm³	b. 11 dm³
c. 7.75 dm³	c. 7.8 cm³
d. 15 dm³	d. 15 cm³
e. 3.5 dm³	e. 56 cm³

Give the mass quantity of the following capacities of water:

Convert to kilograms: Group 73	Convert to grams: Group 74
a. 3 l	a. 9 ml
b. 85 l	b. 45 ml
c. 11 l	c. 10.5 ml
d. 5.25 l	d. 2.25 cl
e. 14.75 l	e. 3 cl

PRACTICE EXAMPLES - VOLUME, CAPACITY, AND MASS (continued)

Exercise 20

Give the capacity occupied by the following mass quantities of water:

Group 75	Group 76
a. 7 kg	a. 6 g
b. 35 kg	b. 15 g
c. 10 kg	c. 72 g
d. 8.25 kg	d. 0.35 g
e. 0.37 kg	e. 30.25 g

Give the volume occupied by the following mass quantities of water:

Group 77	Group 78
a. 5 kg	a. 6 g
b. 12 kg	b. 35 g
c. 90 kg	c. 48 g
d. 2.8 kg	d. 3.2 g
e. 25.67 kg	e. 62.5 g

IX. TEMPERATURE

In practice temperature is expressed in **degrees Celsius** (ºC). The zero point on this scale is the temperature at which water freezes, while the boiling point of water is 100 ºC.

Practical Application

Celsius: Replaces Fahrenheit. Atmospheric and body temperatures are expressed in ºC. Oven temperatures for baking purposes will also be given in ºC.

10 ºC = 50 º F. For every increase or decrease of 5 degrees on the Celsius scale, 9 degrees are added or subtracted from the Fahrenheit scale.

or   Number of degrees Celsius
    = (Number of degrees Fahrenheit — 32) $\times \frac{5}{9}$
    Number of degrees Fahrenheit
    = (Number of degrees Celsius $\times \frac{9}{5}$) + 32

X. CONDENSED TABULATION OF METRIC MEASURES

LIST OF PREFIXES

mega   means a million times
kilo   means a thousand times
hecto   means a hundred times
deka   means ten times
deci   means a tenth part of
centi   means a hundredth part of
milli   means a thousandth part of
micro   means a millionth part of

LENGTH

10 millimetres (mm)   = 1 centimetre (cm)
10 centimetres   = 1 decimetre (dm) = 100 millimetres
10 decimetres   = 1 metre (m) = 1 000 millimetres
10 metres   = 1 dekametre (dam)
10 dekametres   = 1 hectometre (hm) = 100 metres
10 hectometres   = 1 kilometre (km)   = 1 000 metres

1 centimetre (cm)   = 10 millimetres (mm)
1 metre (m)   = 100 centimetres (cm)
                      = 1 000 millimetres (mm)
1 kilometre (km)   = 1 000 metres (m)

AREA

100 square millimetres (mm²)   = 1 square centimetre (cm²)
100 square centimetres (cm²)   = 1 square decimetre (dm²)
100 square decimetres (dm²)   = 1 square metre (m²)
1 000 000 square metres (m²)   = 1 square kilometre (km²)

CONDENSED TABULATION OF METRIC MEASURES
(continued)

CAPACITY (Liquid)

10 millilitres (ml)	= 1 centilitre (cl)	
10 centilitres	= 1 decilitre (dl)	= 100 millilitres
10 decilitres	= 1 litre (l)	= 1 000 millilitres
10 litres	= 1 dekalitre (dal)	
10 dekalitres	= 1 hectolitre (hl)	= 100 litres
10 hectolitres	= 1 kilolitre (kl)	= 1 000 litres

MISCELLANEOUS EQUIVALENTS

1 litre = 1 cubic decimetre (dm^3) = 1 000 cubic centimetres (cm^3)

1 millilitre (ml) = 1 cubic centimetre (cm^3)

1 litre of water weighs 1 kilogram (kg)

1 millilitre (ml) or cubic centimetre (cm^3) of water weighs 1 gram (g)

VOLUME (Dry)

1 000 cubic millimetres (mm^3)	= 1 cubic centimetre (cm^3)	
1 000 cubic centimetres	= 1 cubic decimetre (dm^3)	
	= 1 000 000 cubic millimetres	
1 000 cubic decimetres	= 1 cubic metre (m^3)	
	= 1 000 000 cubic centimetres	
	= 1 000 000 000 cubic millimetres	

CONDENSED TABULATION OF METRIC MEASURES
(continued)

MASS

```
10 milligrams (mg)  =  1 centigram (cg)
10 centigrams       =  1 decigram (dg)  =  100 milligrams
10 decigrams        =  1 gram (g)  =  1 000 milligrams
10 grams            =  1 dekagram (dag)
10 dekagrams        =  1 hectogram (hg)  =  100 grams
10 hectograms       =  1 kilogram (kg)  =  1 000 grams
1 000 kilograms     =  1 metric ton (t)
```

Linear Measure Equivalents

```
1 inch                  = 2.54 centimetres, or 25.4 millimetres.
1 foot                  = 30.4799 centimetres, 304.799
                          millimetres, or 0.3047 metre
1 yard                  = 0.914399 metre
1 mile                  = 1.6093 kilometres  =  5280 feet
1 millimetre            = 0.03937 inch
1 centimetre            = 0.3937 inch
1 decimetre             = 3.937 inches
1 metre                 = 39.370113 inches
                        = 3.28084 feet
                        = 1.093614 yards
1 kilometre             = 0.62137 mile
1 dekametre (10 m)      = 10.936 yards
1 hectometre (100 m)    = 328.084 feet
```

CONDENSED TABULATION OF METRIC MEASURES
(continued)

Metric Conversion Factors

To convert—

millimetres to inches	× 0.03937 or ÷ 25.4
centimetres to inches	× 0.3937 or ÷ 2.54
metres to inches	× 39.37
metres to feet	× 3.281
metres per second to feet per minute	× 197
kilometres to miles	× 0.6214 or ÷ 1.6093
kilometres to feet	× 3280.8693
square millimetres to square inches	× 0.00155 or ÷ 645.1
square centimetres to square inches	× 0.155 or ÷ 6.451
square metres to square feet	× 10.764
square metres to square yards	× 1.2
square kilometres to acres	× 247.1
hectares to acres	× 2.471
cubic centimetres to cubic inches	× 0.06 or ÷ 16.383
cubic metres to cubic feet	× 35.315
cubic metres to cubic yards	× 1.308
cubic metres to gallons (231 cu. in.)	× 164.2
litres to cubic inches	÷ 61.022
litres to gallons	× 0.2642 or ÷ 3.78
litres to cubic feet	× 28.316
hectolitres to cubic feet	× 3.531
hectolitres to bushels (2150.42 cu. in.)	× 2.84
hectolitres to cubic yards	× 0.131

CONDENSED TABULATION OF METRIC MEASURES
(continued)

Metric Conversion Factors

hectolitres to gallons	÷ 26.42
grams to ounces (avoirdupois)	× 0.035 or ÷ 28.35
grams per cubic centimetre to pounds per cubic inch.	÷ 27.7
joules to foot-pounds	× 0.7373
kilograms to ounces	× 35.3
kilograms to pounds	× 2.2046
kilograms to tons	× 0.001
kilograms per square centimetre to pounds per square inch	× 14.223

CONDENSED TABULATION OF METRIC MEASURES
(continued)

Compound Conversion Factors

U.S. - Metric

pounds per lineal foot	×	1.488
	=	kilos per lineal metre
pounds per lineal yard	×	0.496
	=	kilos per lineal metre
tons per lineal foot	×	3333.33
	=	kilos per lineal metre
tons per lineal yard	×	1111.11
	=	kilos per lineal metre
pounds per mile	×	0.2818
	=	kilos per kilometre
pounds per square inch	×	0.0703
	=	kilos per square centimetre
tons per square inch	×	1.575
	=	kilos per square millimetre
pounds per square foot	×	4.883
	=	kilos per square metre
tons per square foot	×	10.936
	=	tons per square metre
tons per square yard	×	1.215
	=	tons per square metre
pounds per cubic yard	×	0.5933
	=	kilos per cubic metre
pounds per cubic foot	×	16.020
	=	kilos per cubic metre
tons per cubic yard	×	1.329
	=	tons per cubic metre
grains per gallon	×	0.01426
	=	grams per litre
pounds per gallon	×	0.09983
	=	kilos per litre

CONDENSED TABULATION OF METRIC MEASURES
(continued)

Compound Conversion Factors

Metric - U.S.

kilos per lineal metre	× 0.672
	= pounds per lineal foot
kilos per lineal metre	× 2.016
	= pounds per lineal yard
kilos per lineal metre	× 0.0003
	= tons per lineal foot
kilos per lineal metre	× 0.0009
	= tons per lineal yard
kilos per kilometre	× 3.548
	= pounds per mile
kilos per square centimetre	× 14.223
	= pounds per square inch
kilos per square millimetre	× 0.635
	= tons per square inch
kilos per square metre	× 0.2048
	= pounds per square foot
tons per square metre	× 0.0914
	= tons per square foot
tons per square metre	× 0.823
	= tons per square yard
kilos per cubic metre	× 1.686
	= pounds per cubic yard
kilos per cubic metre	× 0.0624
	= pounds per cubic foot
tons per cubic metre	× 0.752
	= tons per cubic yard
grams per litre	× 70.12
	= grains per gallon
kilos per litre	× 10.438
	= pounds per gallon

CONDENSED TABULATION OF METRIC MEASURES
(continued)
Table 1 — Linear Measure
Conversion of price per foot to price per metre

Cents per ft	Cents per m	Cents per ft	Cents per m
1	3.281	51	167.323
2	6.562	52	170.604
3	9.843	53	173.885
4	13.123	54	177.165
5	16.404	55	180.446
6	19.685	56	183.727
7	22.966	57	187.008
8	26.247	58	190.289
9	29.528	59	193.570
10	32.808	60	196.850
11	36.089	61	200.131
12	39.370	62	203.412
13	42.651	63	206.693
14	45.932	64	909.974
15	49.213	65	213.255
16	52.493	66	216.535
17	55.774	67	219.816
18	59.055	68	223.097
19	62.336	69	226.378
20	65.617	70	229.659
21	68.898	71	232.940
22	72.179	72	236.220
23	75.459	73	239.501
24	78.740	74	242.782
25	82.021	75	246.063
26	85.302	76	249.344
27	88.583	77	252.625
28	91.864	78	255.906
29	95.144	79	259.186
30	98.425	80	262.467
31	101.706	81	265.748
32	104.987	82	269.029
33	108.268	83	272.310
34	111.549	84	275.591
35	114.829	85	278.871
36	118.110	86	282.152
37	121.391	87	284.433
38	124.672	88	288.714
39	127.953	89	291.995
40	131.234	90	295·276
41	134.514	91	298.556
42	137.795	92	301.837
43	141.076	93	305.118
44	144.357	94	308.399
45	147.638	95	311.680
46	150.919	96	314.961
47	154.199	97	318.241
48	157.480	98	321.522
49	160.761	99	324.803
50	164.042	100	328.084

CONDENSED TABULATION OF METRIC MEASURES
(continued)
Table 2 — Area (square Measure)
Conversion of price per square foot to price per square metre

Cents per ft²	Cents per m²	Cents per ft²	Cents per m²
1	10.764	51	548.959
2	21.528	52	559.723
3	32.292	53	570.487
4	43.056	54	581.251
5	53.820	55	592.015
6	64.583	56	602.779
7	75.347	57	613.543
8	86.111	58	624.307
9	96.875	59	635.071
10	107.639	60	645.835
11	118.403	61	656.599
12	129.167	62	667.362
13	139.931	63	678.126
14	150.695	64	688.890
15	161.459	65	699.654
16	172.223	66	710.418
17	182.986	67	721.182
18	193.750	68	731.946
19	204.514	69	742.710
20	215.278	70	753.474
21	226.042	71	764.238
22	236.806	72	775.002
23	247.570	73	785.765
24	258.334	74	796.529
25	269.098	75	807.293
26	279.862	76	818.057
27	290.626	77	828.821
28	301.389	78	839.585
29	312.153	79	850.349
30	322.917	80	861.113
31	333.681	81	871.877
32	344.445	82	882.641
33	355.209	83	893.405
34	365.973	84	904.168
35	376.737	85	914.932
36	387.501	86	925.696
37	398.265	87	936.460
38	409.029	88	947.224
39	419.793	89	957.988
40	430.556	90	968.752
41	441.320	91	979.516
42	452.084	92	990.280
43	462.848	93	1 001.04
44	473.612	94	1 011.81
45	484.376	95	1 022.57
46	495.140	96	1 033.34
47	505.904	97	1 044.10
48	516.668	98	1 054.86
49	527.432	99	1 065.63
50	538.196	100	1 076.39

CONDENSED TABULATION OF METRIC MEASURES
(continued)

Table 3 — Area (Square Measure)
Conversion of price per square yard to price per square metre

Cents per yd²	Cents per m²	Cents per yd²	Cents per m²
1	1.196	s51	60.996
2	2.392	52	62.192
3	3.588	53	63.388
4	4.784	54	64.584
5	5.980	55	65.780
6	7.176	56	66.975
7	8.372	57	68.171
8	9.568	58	69.367
9	10.764	59	70.563
10	11.960	60	71.759
11	13.156	61	72.955
12	14.352	62	74.151
13	15.548	63	75.347
14	16.744	64	76.543
15	17.940	65	77.739
16	19.136	66	78.935
17	20.332	67	80.131
18	21.528	68	81.327
19	22.724	69	82.523
20	23.920	70	83.719
21	25.116	71	84.915
22	26.312	72	86.111
23	27.508	73	87.307
24	28.704	74	88.503
25	29.900	75	89.699
26	31.096	76	90.895
27	32.292	77	92.091
28	33.488	78	93.287
29	34.684	79	94.483
30	35.880	80	95.679
31	37.076	81	96.875
32	38.272	82	98.071
33	39.468	83	99.267
34	40.664	84	100.463
35	41.860	85	101.659
36	43.056	86	102.855
37	44.252	87	104.051
38	45.448	88	105.247
39	46.644	89	106.443
40	47.840	90	107.639
41	49.036	91	108.835
42	50.232	92	110.031
43	51.428	93	111.227
44	52.624	94	112.423
45	53.820	95	113.619
46	55.016	96	114.815
47	56.212	97	116.011
48	57.408	98	117.207
49	58.604	99	118.403
50	59.800	100	119.599

CONDENSED TABULATION OF METRIC MEASURES
(continued)

Table 4 — Volume (Cubic Measure)
Conversion of price per cubic yard to price per cubic metre

Cents per yd³	Cents per m³	Cents per yd³	Cents per m³
1	1.308	51	66.706
2	2.616	52	68.013
3	3.924	53	69.321
4	5.232	54	70.629
5	6.540	55	71.937
6	7.848	56	73.245
7	9.156	57	74.553
8	10.464	58	75.861
9	11.772	59	77.169
10	13.080	60	78.477
11	14.388	61	79.785
12	15.695	62	81.093
13	17.003	63	82.401
14	18.311	64	83.709
15	19.619	65	85.017
16	20.927	66	86.325
17	22.235	67	87.633
18	23.543	68	88.941
19	24.851	69	90.249
20	26.159	70	91.557
21	27.467	71	92.865
22	28.775	72	94.172
23	30.083	73	95.480
24	31.391	74	96.788
25	32.699	75	98.096
26	34.007	76	99.404
27	35.315	77	100.712
28	36.623	78	102.020
29	37.931	79	103.328
30	39.239	80	104.636
31	40.547	81	105.944
32	41.854	82	107.252
33	43.162	83	108.560
34	44.470	84	109.868
35	45.778	85	111.176
36	47.086	86	112.484
37	48.394	87	113.792
38	49.702	88	115.100
39	51.010	89	116.408
40	52.318	90	117.716
41	53.626	91	119.024
42	54.934	92	120.331
43	56.242	93	121.639
44	57.550	94	122.947
45	58.858	95	124.255
46	60.166	96	125.563
47	61.474	97	126.871
48	62.782	98	128.179
49	64.090	99	129.487
50	65.398	100	130.795

CONDENSED TABULATION OF METRIC MEASURES
(continued)

Table 5 — Mass
Conversion of price per pound to price per kilogram

Cents per lb	Cents per kg	Cents per lb	Cents per kg
1	2.205	51	112.436
2	4.409	52	114.640
3	6.614	53	116.845
4	8.819	54	119.050
5	11.023	55	121.254
6	13.228	56	123.459
7	15.432	57	125.663
8	17.637	58	127.868
9	19.842	59	130.073
10	22.046	60	132.277
11	24.251	61	134.482
12	26.456	62	136.687
13	28.660	63	138.891
14	30.865	64	141.096
15	33.069	65	143.300
16	35.274	66	145.505
17	37.479	67	147.710
18	39.683	68	149.914
19	41.888	69	152.119
20	44.093	70	154.324
21	46.297	71	156.528
22	48.502	72	158.733
23	50.706	73	160.937
24	52.911	74	163.142
25	55.116	75	165.347
26	57.320	76	167.551
27	59.525	77	169.756
28	61.729	78	171.961
29	63.934	79	174.165
30	66.139	80	176.370
31	68.343	81	178.574
32	70.548	82	180.779
33	72.753	83	182.984
34	74.957	84	185.188
35	77.162	85	187.393
36	79.366	86	189.598
37	81.571	87	191.802
38	83.776	88	194.007
39	85.980	89	196.211
40	88.185	90	198.416
41	90.390	91	200.621
42	92.594	92	202.825
43	94.799	93	205.030
44	97.003	94	207.235
45	99.208	95	209.439
46	101.413	96	211.644
47	103.617	97	213.848
48	105.822	98	216.053
49	108.027	99	218.258
50	110.231	100	220.462

ALPHABETICAL CONVERSION TABLE

The fundamental purpose of this chart is to furnish a source of reference
for units, standards, and conversion factors

1 acre $\begin{cases} =160 \text{ square rods.} \\ =4{,}840 \text{ square yards.} \\ =43{,}560 \text{ square feet.} \end{cases}$

1 barrel=7,056 cubic inches.

1 board foot $\begin{cases} =144 \text{ cubic inches.} \\ =2{,}360 \text{ cubic centimetres.} \end{cases}$

1 B. t. u. (British thermal unit) $\begin{cases} =778 \text{ foot pounds.} \\ =0.2930 \text{ international watt hour.} \\ =0.252 \text{ calorie (I. T.).} \end{cases}$

1 bushel $\begin{cases} =2{,}150.42 \text{ cubic inches.} \\ =1\frac{1}{4} \text{ cubic feet, approx.} \end{cases}$

1 calorie (I. T.) $\begin{cases} =1/860 \text{ international watt hours.} \\ =3.97 \times 10^{-3} \text{ B. t. u.} \end{cases}$

1 carat, metric $\begin{cases} =200 \text{ metric milligrams.} \\ =3.0865 \text{ grains.} \end{cases}$

1 centare (square metre) $\begin{cases} =10.764 \text{ square feet.} \\ =1.196 \text{ square yards.} \end{cases}$

1 centimetre=0.3937 inch.

1 chain (engineers) $\begin{cases} =100 \text{ links of 1 foot each.} \\ =30.48 \text{ metres.} \end{cases}$

1 chain (surveyors or Gunters) $\begin{cases} =4 \text{ rods.} \\ =100 \text{ links.} \\ =66 \text{ feet.} \\ =20.1 \text{ metres.} \end{cases}$

1 cheval (French horsepower)=0.986 horsepower.

1 circular mil $\begin{cases} =\text{Area of circle whose diameter is 1 mil, or } 1/1000 \text{ inch.} \\ =0.000000785 \text{ square inch.} \end{cases}$

1 cord $\begin{cases} =128 \text{ cubic feet.} \\ =3.625 \text{ cubic metres.} \end{cases}$

1 cubic foot $\begin{cases} =1{,}728 \text{ cubic inches.} \\ =60 \text{ pints.} \\ =0.8 \text{ bushel.} \\ =1{,}000 \text{ ounces of water, approx.} \\ =0.028 \text{ cubic metre.} \\ =28.32 \text{ litres.} \end{cases}$

1 cubic foot of water $\begin{cases} =62.4 \text{ pounds.} \\ =1{,}000 \text{ ounces, approx.} \end{cases}$

1 cubic inch=16.39 cubic centimetres.

1 cubic metre $\begin{cases} =35.314 \text{ cubic feet.} \\ =1.308 \text{ cubic yards.} \end{cases}$

1 cubic yard $\begin{cases} =27 \text{ cubic feet.} \\ =0.765 \text{ cubic metre.} \end{cases}$

1 decimetre=3.937 inches.

ALPHABETICAL CONVERSION TABLE—continued

1 dram (fluid) $\begin{cases} =60 \text{ minims.} \\ =3.697 \text{ millilitres.} \\ =4 \text{ cubic centimetres, approx.} \end{cases}$

1 em, 1 pica (printing industry)=1/6 of an inch.

1 fathom (nautical) $\begin{cases} =6 \text{ feet.} \\ =1.83 \text{ metres.} \end{cases}$

1 fluid ounce $\begin{cases} =8 \text{ fluid drams.} \\ =29.573 \text{ millilitres.} \end{cases}$

1 foot $\begin{cases} =12 \text{ inches.} \\ =0.305 \text{ metre.} \end{cases}$

1 foot pound=0.1383 kilogrammetre.

1 furlong (British) $\begin{cases} =220 \text{ yards.} \\ =201.2 \text{ metres.} \end{cases}$

1 gallon (U.S.) $\begin{cases} =231 \text{ cubic inches.} \\ =4 \text{ quarts.} \\ =8 \text{ pints.} \\ =3.875 \text{ litres.} \\ =128 \text{ fluid ounces.} \end{cases}$

1 gallon of water=8.33 pounds at 62° F. (16.67° C.) in air.

1 gallon per cubic foot=133.7 litres per cubic metre.

Gallon (British Imperial and Canadian). $\begin{cases} =277.4 \text{ cubic inches.} \\ =1.201 \text{ U.S. gallons.} \\ =\text{volume of 10 pounds water at 62°} \\ \quad \text{F. (16.67° C.).} \\ =4.546 \text{ litres.} \end{cases}$

1 gill=¼ pint.

1 grain $\begin{cases} =1/7000 \text{ pound avoirdupois.} \\ =0.0648 \text{ gram.} \end{cases}$

1 gram $\begin{cases} =15.43 \text{ grains.} \\ =0.0353 \text{ ounce.} \\ =0.0022 \text{ pound.} \end{cases}$

1 hand=4 inches.

1 hectare (square hectometre)=2.47 acres.

1 horsepower $\begin{cases} =33,000 \text{ foot-pounds per minute.} \\ =42.41 \text{ B. t. u. per minute.} \\ =1.014 \text{ chevals.} \\ =746 \text{ watts.} \end{cases}$

1 hundredweight (British) $\begin{cases} =112 \text{ pounds.} \\ =50.80 \text{ kilograms.} \end{cases}$

1 inch=25.4 millimetres.

1 kilogram $\begin{cases} =2.2046 \text{ pounds.} \\ =35.274 \text{ ounces.} \\ =15432.36 \text{ grains.} \\ =0.0011 \text{ short ton.} \\ =0.00098 \text{ long ton.} \end{cases}$

1 kilometre $\begin{cases} =1000 \text{ metres.} \\ =0.621 \text{ mile.} \end{cases}$

1 kilowatt $\begin{cases} =1.34 \text{ horsepower.} \\ =56.9 \text{ B. t. u. per minute.} \end{cases}$

1 knot (nautical, speed)$\begin{cases}=6080.20 \text{ feet per hour.} \\ =1.85 \text{ kilometres per hour.}\end{cases}$

1 light year $\begin{cases}=5.9\times10^{12} \text{ miles.} \\ =9.5\times10^{12} \text{ kilometres.}\end{cases}$

1 link (surveyors measure)$\begin{cases}=0.66 \text{ foot.} \\ =0.201 \text{ metre.}\end{cases}$

1 litre $\begin{cases}=1.000028 \text{ cubic decimetres.} \\ =0.264 \text{ gallon.} \\ =1.057 \text{ quarts.} \\ =61.03 \text{ cubic inches.} \\ =0.035 \text{ cubic feet.} \\ =33.8148 \text{ fluid ounces.} \\ =270.518 \text{ fluid drams.}\end{cases}$

1 metre $\begin{cases}=39.37 \text{ inches.} \\ =3.28 \text{ feet.} \\ =1.09 \text{ yards.} \\ =1\ 650\ 763.73 \text{ wave lengths in a vacuum of} \\ \quad \text{orange-red radiation of krypton 86.}\end{cases}$

1 metric ton $\begin{cases}=2204.6 \text{ pounds.} \\ =1.1023 \text{ short tons.}\end{cases}$

1 microgram=1/1000 milligram.

1 mil$\begin{cases}=0.001 \text{ inch.} \\ =25.4 \text{ microns.} \\ =0.0254 \text{ millimetre.}\end{cases}$

1 mile $\begin{cases}=1760 \text{ yards.} \\ =5280 \text{ feet.} \\ =320 \text{ rods.} \\ =1.61 \text{ kilometres.}\end{cases}$

1 milligram=0.0154 grain.

1 millilitre (see litre above)$\begin{cases}=1.000028 \text{ cubic centimetres.} \\ =0.0610 \text{ cubic inch.}\end{cases}$

1 minim (fluid)$\begin{cases}=1/60 \text{ fluid dram.} \\ =1/480 \text{ fluid ounce.}\end{cases}$

1 ounce (avoirdupois, ordinary)$\begin{cases}=437.5 \text{ grains.} \\ =0.911 \text{ troy ounce.} \\ =0.0000279 \text{ long ton.} \\ =28.35 \text{ grams.}\end{cases}$

1 ounce, fluid$\begin{cases}=1.805 \text{ cubic inches.} \\ =29.573 \text{ millilitres.}\end{cases}$

1 ounce, troy$\begin{cases}=480 \text{ grains.} \\ =31.103 \text{ grams.}\end{cases}$

1 perch (British)$\begin{cases}=30.25 \text{ square yards.} \\ =1/160 \text{ acre.}\end{cases}$

1 pied (French foot)$\begin{cases}=12 \text{ Paris inches.} \\ =0.325 \text{ metre.}\end{cases}$

1 pint=0.4732 litre.

1 point (printers type)=1/72 inch.

1 pole (British)$\begin{cases}=5\frac{1}{2} \text{ yards.} \\ =5.03 \text{ metre.} \\ =1 \text{ rod.}\end{cases}$

ALPHABETICAL CONVERSION TABLE—continued

1 pouce (Paris inch)=2.71 centimetre.

1 pound (avoirdupois, ordinary) {=16 ounces. =7000 grains. =454 grams. =0.454 kilogram. =14.58 troy ounces.

1 pound per cubic foot=16.02 kilogram per cubic metre.

1 pound per square inch=0.433×head of water (in feet).

1 pound per square inch=0.0703 kilogram per square centimetre.

1 pound per square foot=4.88 kilogram per square metre.

1 quart {=2 pints. =¼ gallon. =0.946 litre.

1 quarter (British quarter hundredweight) {=28 pounds. =12.70 kilograms.

1 rod (surveyor's measure) {=16.5 feet. =25 links. =5.03 metres.

1 rood (British) {=40 perches. =¼ acre.

1 square centimetre=0.155 square inch.

1 square foot=0.093 square metre.

1 square inch=6.452 square centimetres.

1 square kilometre=0.386 square mile.

1 square metre (centare) {=10.764 square feet. =1.196 square yards.

1 square mil {=0.000001 square inch. =0.00000645 square centimetre.

1 square mile {=640 acres. =3,097,600 square yards. =2.59 square kilometres.

1 square millimetre=0.00155 square inch.

1 square rod=25.29 square metres.

1 square yard=0.836 square metre.

1 stone (British) {=14 pounds. =6.35 kilograms.

1 ton (short) {=2,000 pounds. =907 kilograms.

1 ton (long) {=2,240 pounds. =1,016 kilograms.

1 ton (metric) {=1,000 kilograms. =2,204.62 pounds.

1 yard {=3 feet. =36 inches. =0.914 metre. =1 508 798.05 wave lengths in a vacuum of orange-red radiation of krypton 86.

> "Never has anything more grand and more simple, more coherent in all its parts, issued from the hand of men."
> — Antoine Laurent Lavoisier
>   (1743-1794)

SI UNITS, MULTIPLES, SUBMULTIPLES AND PREFIXES

ampere	A	are	a
ampere metre squared	A.m²	atto-	a
ampere per metre	A/m	bar	bar
ampere per millimetre	A/mm	candela	cd
ampere per square centimetre	A/cm²	candela per square metre	cd/m²
ampere per square metre	A/m²	centi-	c
		centimetre	cm
ampere per square millimetre	A/mm²	centipoise	cP
		centistokes	cSt
		coulomb	C

coulomb metre	C.m	gram per cubic	
coulomb per cubic		centimetre	g/cm³
centimetre	C/cm³	gram per litre	g/l
coulomb per cubic		gram per millilitre	g/ml
metre	C/m³	gram per mole	g/mol
coulomb per cubic		hectare	ha
millimetre	C/mm³	hecto-	h
coulomb per square		hectobar	hbar
centimetre	C/cm²	henry	H
coulomb per square		henry per metre	H/m
metre	C/m²	hertz	Hz
coulomb per square		hour	hr
millimetre	C/mm²	joule	J
cubic centimetre	cm³	joule per degree	
cubic decimetre	dm³	celsius	J/°C
cubic metre	m³	joule per kelvin	J/K
cubic millimetre	mm³	joule per kilogram	J/kg
cubic metre per		joule per kilogram	
kilomole	m³/kmol	degree celsius	J/kg.°C
cubic metre per		joule per kilogram	
mole	m³/mol	kelvin	J/kg.K
day	d	joule per kilomole	J/kmol
deca-	da	joule per kilomole	
decanewton	daN	degree celsius	J/kmol.°C
deci-	d	joule per kilomole	
decimetre	dm	kelvin	J/kmol.K
degree (angles)	…°	joule per mole	J/mol
degree celsius	°C	joule per mole	
farad	F	degree celsius	J/mol.°C
farad per metre	F/m	joule per mole	
femto-	f	kelvin	J/mol.K
giga-	G	joule per square	
gigahertz	GHz	centimetre	J/cm²
gigajoule	GJ	joule per square	
giganewton per		metre	J/m²
square metre	GN/m²	kelvin	K
gigaohm	GΩ	kilo-	k
gigaohm metre	GΩ.m	kiloampere	kA
gigawatt	GW	kiloampere per	
gigawatt hour	GW.h	metre	kA/m
gram	g	kiloampere per	
		square metre	kA/m²

kilocoulomb	kC
kilocoulomb per cubic metre	kC/m³
kilocoulomb per square metre	kC/m²
kilogram	kg
kilogram metre per second	kg.m/s
kilogram metre squared per second	kg.m²/s
kilogram per cubic decimetre	kg/dm³
kilogram per cubic metre	kg/m³
kilogram per litre	kg/*l*
kilogram per mole	kg/mo*l*
kilohertz	kHz
kilojoule	kJ
kilojoule per degree celsius	kJ/°C
kilojoule per kelvin	kJ/K
kilojoule per kilogram	kJ/kg
kilojoule per kilogram degree celsius	kJ/kg.°C
kilojoule per kilogram kelvin	kJ/kg.K
kilojoule per square metre	kJ/m²
kilolitre	k*l*
kilometre	km
kilometre per hour	km/h
kilomole	kmo*l*
kilomole per cubic metre	kmo*l*/m³
kilomole per kilogram	kmo*l*/kg
kilomole per litre	kmo*l*/*l*
kilonewton	kN
kilonewton metre	kN.m

kilonewton per square metre	kN/m²
kilo-ohm	kΩ
kilo-ohm metre	kΩ.m
kilosecond	ks
kilosiemens	kS
kilosiemens per metre	kS/m
kilovolt	kV
kilovolt per metre	kV/m
kilowatt	kW
kilowatt hour	kW.h
kilowatt per square metre	kW/m²
kiloweber per metre	kWb/m
litre	*l*
lumen	*l*m
lumen hour	*l*m.h
lumen second	*l*m.s
lux	*l*x
mega-	M
mega-ampere per square metre	MA/m²
megacoulomb	MC
megacoulomb per cubic metre	MC/m³
megacoulomb per square metre	MC/m²
megagram	Mg
megagram per cubic metre	Mg/m³
megahertz	MHz
megajoule	MJ
megajoule per kilogram	MJ/kg
meganewton	MN
meganewton metre	MN.m
meganewton per square metre	MN/m²
megaohm	MΩ
megaohm metre	MΩ.m
megasiemens per metre	MS/m

59

megavolt	MV	milliampere	mA
megavolt per metre	MV/m	millibar	mbar
megawatt	MW	millicoulomb	mC
megawatt hour	MW.h	millifarad	mF
megawatt per square		millifarad per metre	mF/m
metre	MW/m^2	milligram	mg
metre	m	millihenry	mH
metre per second	m/s	millijoule	mJ
metre per second		millilitre	m*l*
squared	m/s^2	millimetre	mm
metre raised to the		millimetre squared	
fourth power	m^4	per second	mm^2/s
metre squared per		millinewton	mN
second	m^2/s	millinewton per	
metric ton	t	metre	mN/m
metric ton per cubic		millinewton per	
metre	t/m^3	square metre	mN/m^2
micro-	μ	millinewton second	
microampere	μA	per metre	
microbar	μbar	squared	mN.s/m^2
microcoulomb	μC	milliohm	mΩ
microfarad	μF	milliohm metre	mΩ.m
microfarad per metre	μF/m	milliradian	mrad
microgram	μg	millisecond	ms
microhenry	μH	millisiemens	mS
microhenry per		millitesla	mT
metre	μH/m	millivolt	mV
micrometre	μm	millivolt per metre	mV/m
micronewton	μN	milliwatt	mW
micronewton metre	μN.m	milliweber	mWb
micronewton per		minute (angles)	. . .$'$
square metre	μN/m^2	minute	min
micro-ohm	$\mu\Omega$	mole	mo*l*
micro-ohm metre	$\mu\Omega$.m	mole per cubic	
microradian	μrad	metre	mo*l*/m^3
microsecond	μs	mole per kilogram	mo*l*/kg
microsiemens	μS	mole per litre	mo*l*/*l*
microtesla	μT	nano-	n
microvolt	μV	nanoampere	nA
microvolt per metre	μV/m	nanocoulomb	nC
microwatt	μW	nanofarad	nF
milli-	m	nanofarad per metre	nF/m

60

nanohenry	nH	second (angles)	$\ldots''$
nanohenry per metre	nH/m	siemens	S
nanometre	nm	siemens per metre	S/m
nano-ohm metre	$n\Omega.m$	square centimetre	cm^2
nanosecond	ns	square decimetre	dm^2
nanotesla	nT	square kilometre	km^2
nanowatt	nW	square metre	m^2
newton	N	square millimetre	mm^2
newton metre	N.m	steradian	sr
newton metre		tera-	T
squared per		terahertz	THz
ampere	$N.m^2/A$	terajoule	TJ
newton per metre	N/m	terawatt	TW
newton per square		tesla	T
metre	N/m^2	volt	V
newton per square		volt ampere	V.A
millimetre	N/mm^2	volt per centimetre	V/cm
newton second per		volt per metre	V/m
metre squared	$N.s/m^2$	volt per millimetre	V/mm
ohm	Ω	volt second	V.s
ohm metre	$\Omega.m$	watt	W
ohm millimetre	$\Omega.mm$	watt hour	W.h
pascal	Pa	watt per metre	
per degree celsius	$°C^{-1}$	degree celsius	W/m.°C
per henry	H^{-1}	watt per metre	
per kelvin	K^{-1}	kelvin	W/m.K
per ohm	Ω^{-1}	watt per square	
per ohm metre	$\Omega^{-1}.m^{-1}$	metre	W/m^2
per second	s^{-1}	watt per square	
pico-	p	metre degree	
picoampere	pA	celsius	$W/m^2.°C$
picocoulomb	pC	watt per square	
picofarad	pF	metre kelvin	$W/m.^2K$
picofarad per metre	pF/m	watt per steradian	W/sr
picohenry	pH	weber	Wb
radian	rad	weber metre	Wb.m
radian per second	rad/s	weber per metre	Wb/m
radian per second		weber per milli-	
squared	rad/s^2	metre	Wb/mm
second	s		

TABLES OF INTERRELATION
1. UNITS OF

Units	Inches	Links	Feet	Yards	Rods
1 inch =	1	0.126 263	0.083 333 3	0.027 777 8	0.005 050 51
1 link =	7.92	1	0.66	0.22	0.04
1 foot =	12	1.515 152	1	0.333 333	0.060 606 1
1 yard =	36	4.545 45	3	1	0.181 818
1 rod =	198	25	16.5	5.5	1
1 chain =	792	100	66	22	4
1 mile =	63 360	8000	5280	1760	320
1 centimetre =	0.3937	0.049 709 60	0.032 808 33	0.010 936 111	0.001 988 384
1 metre =	39.37	4.970 960	3.280 833	1.093 611 1	0.198 838 4

2. UNITS OF

Units	Square inches	Square links	Square feet	Square yards	Square rods	Square chains
1 square inch =	1	0.015 942 3	0.006 944 44	0.000 771 605	0.000 025 507 6	0.000 001 594 23
1 square link =	62.7264	1	0.4356	0.0484	0.0016	0.0001
1 square foot =	144	2.295 684	1	0.111 111 1	0.003 673 09	0.000 229 568
1 square yard =	1296	20.6612	9	1	0.033 057 85	0.002 066 12
1 square rod =	39 204	625	272.25	30.25	1	0.0625
1 square chain =	627 264	10 000	4356	484	16	1
1 acre =	6 272 640	100 000	43 560	4840	160	10
1 square mile =	4 014 489 600	64 000 000	27 878 400	3 097 600	102 400	6400
1 square centimetre =	0.154 999 69	0.002 471 04	0.001 076 387	0.000 119 598 5	0.000 003 953 67	0.000 000 247 104
1 square metre =	1549.9969	24.7104	10.763 87	1.195 985	0.039 536 7	0.002 471 04
1 hectare =	15 499 969	247 104	107 638.7	11 959.85	395.367	24.7104

3. UNITS OF

Units	Cubic inches	Cubic feet	Cubic yards
1 cubic inch =	1	0.000 578 704	0.000 021 433 47
1 cubic foot =	1728	1	0.037 037 0
1 cubic yard =	46 656	27	1
1 cubic centimetre =	0.061 023 38	0.000 035 314 45	0.000 001 307 94
1 cubic decimetre =	61.023·38	0.035 314 45	0.001 307 943
1 cubic metre =	61 023.38	35.314 45	1.307 942 8

4. UNITS OF CAPACITY

Units	Minims	Fluid drams	Fluid ounces	Gills	Liquid pints
1 minim =	1	0.016 666 7	0.002 083 33	0.000 520 833	0.000 130 208
1 fluid dram =	60	1	0.125	0.031 25	0.007 812 5
1 fluid ounce =	480	8	1	0.25	0.0625
1 gill =	1920	32	4	1	0.25
1 liquid pint =	7680	128	16	4	1
1 liquid quart =	15 360	256	32	8	2
1 gallon =	61 440	1024	128	32	8
1 millilitres =	16.2311	0.270 518	0.033 814 7	0.008 453 68	0.002 113 42
1 liter =	16 231.1	270.518	33.8147	8.453 68	2.113 42
1 cubic inch =	265.974	4.432 90	0.554 113	0.138 528	0.034 632 0

OF UNITS OF MEASUREMENT
LENGTH

Chains	Miles	Centimetres	Metres	Units
0.001 262 63	0.000 015 782 8	2.540 005	0.025 400 05	=1 inch
0.01	0.000 125	20.116 84	0.201 168 4	=1 link
0.015 151 5	0.000 189 393 9	30.480 06	0.304 800 6	=1 foot
0.045 454 5	0.000 568 182	91.440 18	0.914 401 8	=1 yard
0.25	0.003 125	502.9210	5.029 210	=1 rod
1	0.0125	2011.684	20.116 84	=1 chain
80	1	160 934.72	1609.3472	=1 mile
0.000 497 096 0	0.000 006 213 699	1	0.01	=1 centimetre
0.049 709 60	0.000 621 369 9	100	1	=1 metre

AREA

Acres	Square miles	Square centimetres	Square metres	Hectares	Units
0.000 000 159 423	0.000 000 000 249 1	6.451 626	0.000 645 162 6	0.000 000 064 516	=1 square inch
0.000 01	0.000 000 015 625	404.6873	0.040 468 73	0.000 004 046 87	=1 square link
0.000 022 956 8	0.000 000 035 870 1	929.0341	0.092 903 41	0.000 009 290 34	=1 square foot
0.000 206 612	0.000 000 322 831	8361.307	0.836 130 7	0.000 083 613 1	=1 square yard
0.006 25	0.000 009 765 625	252 929.5	25.292 95	0.002 529 295	=1 square rod
0.1	0.000 156 25	4 046 873	404.6873	0.040 468 7	=1 square chain
1	0.001 562 5	40 468 726	4046.873	0.404 687	=1 acre
640	1	25 899 984 703	2 589 998	258.9998	=1 square mile
0.000 000 024 710 4	0.000 000 000 038 610 06	1	0.0001	0.000 000 01	=1 square centimetre
0.000 247 104	0.000 000 386 100 6	10 000	1	0.0001	=1 square metre
2.471 04	0.003 861 006	100 000 000	10 000	1	=1 hectare

VOLUME

Cubic centimetres	Cubic decimetres	Cubic metres	Units
16.387 162	0.016 387 16	0.000 016 387 16	=1 cubic inch
28 317.016	28.317 016	0.028 317 016	=1 cubic foot
764 559.4	764.5594	0.764 559 4	=1 cubic yard
1	0.001	0.000 001	=1 cubic centimetre
1 000	1	0.001	=1 cubic decimetre
1 000 000	1000	1	=1 cubic metre

LIQUID MEASURE

Liquid quarts	Gallons	Milliliters	Liters	Cubic inches	Units
0.000 065 104	0.000 016 276	0.061 610 2	0.000 061 610 2	0.003 759 77	=1 minim
0.003 906 25	0.000 976 562	3.696 61	0.003 696 61	0.225 586	=1 fluid dram
0.031 25	0.007 812 5	29.5729	0.029 572 9	1.804 69	=1 fluid ounce
0.125	0.031 25	118.292	0.118 292	7.218 75	=1 gill
0.5	0.125	473.167	0.473 167	28.875	=1 liquid pint
1	0.25	946.333	0.946 333	57.75	=1 liquid quart
4	1	3785.332	3.785 332	231	=1 gallon
0.001 056 71	0.000 264 178	1	0.001	0.061 025 0	=1 milliliter
1.056 71	0.264 178	1000	1	61.0250	=1 liter
0.017 316 0	0.004 329 00	16.3867	0.016 386 7	1	=1 cubic inch

TABLES OF INTERRELATION
5. UNITS OF CAPACITY

Units	Dry pints	Dry quarts	Pecks	Bushels
1 dry pint =	1	0.5	0.0625	0.015 625
1 dry quart =	2	1	0.125	0.031 25
1 peck =	16	8	1	0.25
1 bushel =	64	32	4	1
1 liter =	1.816 20	0.908 102	0.113 513	0.028 378
1 dekalitre =	18.1620	9.081 02	1.135 13	0.283 78
1 cubic inch=	0.029 761 6	0.014 880 8	0.001 860 10	0.000 465 025

6. UNITS OF MASS LESS

Units *	Grains	Apothecaries' scruples	Pennyweights	Avoirdupois drams	Apothecaries' drams	Avoirdupois ounces
1 grain =	1	0.05	0.041 666 67	0.036 571 43	0.016 666 7	0.002 285 71
1 apoth. scruple =	20	1	0.833 333 3	0.731 428 6	0.333 333	0.045 714 3
1 pennyweight =	24	1.2	1	0.877 714 3	0.4	0.054 857 1
1 avoir. dram =	27.343 75	1.367 187 5	1.139 323	1	0.455 729 2	0.0625
1 apoth. dram =	60	3	2.5	2.194 286	1	0.137 142 9
1 avoir. ounce =	437.5	21.875	18.229 17	16	7.291 67	1
1 apoth. or troy ounce=	480	24	20	17.554 28	8	1.097 142 9
1 apoth. or troy pound=	5760	288	240	210.6514	96	13.165 714
1 avoir. pound =	7000	350	291.6667	256	116.6667	16
1 milligram =	0.015 432 356	0.000 771 618	0.000 643 014 8	0.000 564 383 3	0.000 257 205 9	0.000 035 273 96
1 gram =	15.432 356	0.771 618	0.643 014 85	0.564 383 3	0.257 205 9	0.035 273 96
1 kilogram =	15 432.356	771.6178	643.014 85	564.383 32	257.205 94	35.273 96

7. UNITS OF MASS

Units	Avoirdupois ounces	Avoirdupois pounds	Short hundred-weights	Short tons
1 avoirdupois ounce =	1	0.0625	0.000 625	0.000 031 25
1 avoirdupois pound =	16	1	0.01	0.0005
1 short hundredweight=	1600	100	1	0.05
1 short ton =	32 000	2000	20	1
1 long ton =	35 840	2240	22.4	1.12
1 kilogram =	35.273 957	2.204 622 34	0.022 046 223	0.001 102 311.2
1 metric ton =	35 273.957	2204.622 34	22.046 223	1.102 311 2

* "Avoir." is now abbreviated "avdp".

OF UNITS OF MEASUREMENT
DRY MEASURE

Litres	Dekalitres	Cubic inches	Units
0.550 599	0.055 060	33.600 312 5	=1 dry pint
1.101 198	0.110 120	67.200 625	=1 dry quart
8.809 58	0.880 958	537.605	=1 peck
35.2383	3.523 83	2150.42	=1 bushel
1	0.1	61.0250	=1 liter
10	1	610.250	=1 dekalitre
0.016 386 7	0.001 638 67	1	=1 cubic inch

THAN POUNDS AND KILOGRAMS

Apothecaries' or troy ounces	Apothecaries' or troy pounds	Avoirdupois pounds	Milligrams	Grams	Kilograms	Units
0.002 083 33	0.000 173 611 1	0.000 142 857 1	64.798 918	0.064 798 918	0.000 064 798 9	=1 grain
0.041 666 7	0.003 472 222	0.002 857 143	1295.9784	1.295 978 4	0.001 295 978	=1 apoth. scruple
0.05	0.004 166 667	0.003 428 571	1555.1740	1.555 174 0	0.001 555 174	=1 pennyweight
0.056 966 146	0.004 747 178 8	0.003 906 25	1771.8454	1.771 845 4	0.001 771 845	=1 avoir. dram
0.125	0.010 416 667	0.008 571 429	3887.9351	3.887 935 1	0.003 887 935	=1 apoth. dram
0.911 458 3	0.075 954 861	0.0625	28 349.527	28.349 527	0.028 349 53	=1 avoir. ounce
1	0.083 333 33	0.068 571 43	31 103.481	31.103 481	0.031 103 48	=1 apoth. or troy ounce
12	1	0.822 857 1	373 241.77	373.241 77	0.373 241 77	=1 apoth. or troy pound
14.583 333	1.215 277 8	1	453 592.4277	453.592 4277	0.453 592 427 7	=1 avoir. pound
0.000 032 150 74	0.000 002 679 23	0.000 002 204 62	1	0.001	0.000 001	=1 milligram
0.032 150 74	0.002 679 23	0.002 204 62	1000	1	0.001	=1 gram
32.150 742	2.679 228 5	2.204 622 341	1 000 000	1000	1	=1 kilogram

GREATER THAN AVOIRDUPOIS OUNCES

Long tons	Kilograms	Metric tons	Units
0.000 027 901 79	0.028 349 53	0.000 028 349 53	=1 avoirdupois ounce
0.000 446 428 6	0.453 592 427 7	0.000 453 592 43	=1 avoirdupois pound
0.044 642 86	45.359 243	0.045 359 243	=1 short hundredweight
0.892 857 1	907.184 86	0.907 184 86	=1 short ton
1	1016.047 04	1.016 047 04	=1 long ton
0.000 984 206 4	1	0.001	=1 kilogram
0.984 206 40	1000	1	=1 metric ton

COMPARISON OF METRIC AND CUSTOMARY UNITS FROM 1 TO 9

1. LENGTH

Inches (in.)	Milli-metres (mm)	Feet (ft)	Metres (m)	Yards (yd)	Metres (m)	Rods (rd)	Metres (m)	U.S. miles (mi)	Kilometres (km)
1= 25.4001		1=0.304 801		1=0.914 402		1= 5.029 21		1= 1.609 347	
2= 50.8001		2=0.609 601		2=1.828 804		2=10.058 42		2= 3.218 694	
3= 76.2002		3=0.914 402		3=2.743 205		3=15.087 63		3= 4.828 042	
4=101.6002		4=1.219 202		4=3.657 607		4=20.116 84		4= 6.437 389	
5=127.0003		5=1.524 003		5=4.572 009		5=25.146 05		5= 8.046 736	
6=152.4003		6=1.828 804		6=5.486 411		6=30.175 26		6= 9.656 083	
7=177.8004		7=2.133 604		7=6.400 813		7=35.204 47		7=11.265 431	
8=203.2004		8=2.438 405		8=7.315 215		8=40.233 68		8=12.874 778	
9=228.6005		9=2.743 205		9=8.229 616		9=45.262 89		9=14.484 125	
0.039 37=1		3.280 83=1		1.093 611=1		0.198 838=1		0.621 370=1	
0.078 74=2		6.561 67=2		2.187 222=2		0.397 677=2		1.242 740=2	
0.118 11=3		9.842 50=3		3.280 833=3		0.596 515=3		1.864 110=3	
0.157 48=4		13.123 33=4		4.374 444=4		0.795 354=4		2.485 480=4	
0.196 85=5		16.404 17=5		5.468 056=5		0.994 192=5		3.106 850=5	
0.236 22=6		19.685 00=6		6.561 667=6		1.193 030=6		3.728 220=6	
0.275 59=7		22.965 83=7		7.655 278=7		1.391 869=7		4.349 590=7	
0.314 96=8		26.246 67=8		8.748 889=8		1.590 707=8		4.970 960=8	
0.354 33=9		29.527 50=9		9.842 500=9		1.789 545=9		5.592 330=9	

2. AREA

Square inches (sq in.)	Square centimetres (cm²)	Square feet (sq ft)	Square metres (m²)	Square yards (sq yd)	Square metres (m²)	Acres (acre)	Hectares (ha)	Square miles (sq mi)	Square Kilometres (km²)
1= 6.452		1=0.092 90		1=0.8361		1=0.4047		1= 2.5900	
2=12.903		2=0.185 81		2=1.6723		2=0.8094		2= 5.1800	
3=19.355		3=0.278 71		3=2.5084		3=1.2141		3= 7.7700	
4=25.807		4=0.371 61		4=3.3445		4=1.6187		4=10.3600	
5=32.258		5=0.464 52		5=4.1807		5=2.0234		5=12.9500	
6=38.710		6=0.557 42		6=5.0168		6=2.4281		6=15.5400	
7=45.161		7=0.650 32		7=5.8529		7=2.8328		7=18.1300	
8=51.613		8=0.743 23		8=6.6890		8=3.2375		8=20.7200	
9=58.065		9=0.836 13		9=7.5252		9=3.6422		9=23.3100	
0.155 00=1		10.764=1		1.1960=1		2.471=1		0.3861=1	
0.310 00=2		21.528=2		2.3920=2		4.942=2		0.7722=2	
0.465 00=3		32.292=3		3.5880=3		7.413=3		1.1583=3	
0.620 00=4		43.055=4		4.7839=4		9.884=4		1.5444=4	
0.775 00=5		53.819=5		5.9799=5		12.355=5		1.9305=5	
0.930 00=6		64.583=6		7.1759=6		14.826=6		2.3166=6	
1.085 00=7		75.347=7		8.3719=7		17.297=7		2.7027=7	
1.240 00=8		86.111=8		9.5679=8		19.768=8		3.0888=8	
1.395 00=9		96.875=9		10.7639=9		22.239=9		3.4749=9	

3. VOLUME

Cubic inches (cu in.)	Cubic centimetres (cm³)	Cubic feet (cu ft)	Cubic metres (m³)	Cubic yards (cu yd)	Cubic metres (m³)	Cubic inches (cu in.)	Litres (litre)	Cubic feet (cu ft)	Litres (litre)
1= 16.3872		1=0.028 317		1=0.7646		1=0.016 386 7		1= 28.316	
2= 32.7743		2=0.056 634		2=1.5291		2=0.032 773 4		2= 56.633	
3= 49.1615		3=0.084 951		3=2.2937		3=0.049 160 2		3= 84.949	
4= 65.5486		4=0.113 268		4=3.0582		4=0.065 546 9		4=113.265	
5= 81.9358		5=0.141 585		5=3.8228		5=0.081 933 6		5=141.581	
6= 98.3230		6=0.169 902		6=4.5874		6=0.098 320 3		6=169.898	
7=114.7101		7=0.198 219		7=5.3519		7=0.114 707 0		7=198.214	
8=131.0973		8=0.226 536		8=6.1165		8=0.131 093 8		8=226.530	
9=147.4845		9=0.254 853		9=6.8810		9=0.147 480 5		9=254.846	
0.061 02=1		35.314=1		1.3079=1		61.025=1		0.035 315=1	
0.122 05=2		70.629=2		2.6159=2		122.050=2		0.070 631=2	
0.183 07=3		105.943=3		3.9238=3		183.075=3		0.105 946=3	
0.244 09=4		141.258=4		5.2318=4		244.100=4		0.141 262=4	
0.305 12=5		176.572=5		6.5397=5		305.125=5		0.176 577=5	
0.366 14=6		211.887=6		7.8477=6		366.150=6		0.211 892=6	
0.427 16=7		247.201=7		9.1556=7		427.175=7		0.247 208=7	
0.488 19=8		282.516=8		10.4635=8		488.200=8		0.282 523=8	
0.549 21=9		317.830=9		11.7715=9		549.225=9		0.317 839=9	

4. CAPACITY—LIQUID MEASURE

U. S. fluid drams (fl dr)	Millilitres (ml)	U. S. fluid ounces (fl oz)	Millilitres (ml)	U.S. liquid pints (pt)	Litres (litre)	U.S. liquid quarts (qt)	Litres (litre)	U. S. gallons (gal)	Litres (litre)
1= 3.6966		1= 29.573		1=0.473 17		1=0.946 33		1= 3.785 33	
2= 7.3932		2= 59.146		2=0.946 33		2=1.892 67		2= 7.570 66	
3=11.0898		3= 88.719		3=1.419 50		3=2.839 00		3=11.356 00	
4=14.7865		4=118.292		4=1.892 67		4=3.785 33		4=15.141 33	
5=18.4831		5=147.865		5=2.365 83		5=4.731 67		5=18.926 66	
6=22.1797		6=177.437		6=2.839 00		6=5.678 00		6=22.711 99	
7=25.8763		7=207.010		7=3.312 17		7=6.624 33		7=26.497 33	
8=29.5729		8=236.583		8=3.785 33		8=7.570 66		8=30.282 66	
9=33.2695		9=266.156		9=4.258 50		9=8.517 00		9=34.067 99	
0.270 52=1		0.033 815=1		2.1134=1		1.056 71=1		0.264 18=1	
0.541 04=2		0.067 629=2		4.2268=2		2.113 42=2		0.528 36=2	
0.811 55=3		0.101 444=3		6.3403=3		3.170 13=3		0.792 53=3	
1.082 07=4		0.135 259=4		8.4537=4		4.226 84=4		1.056 71=4	
1.352 59=5		0.169 074=5		10.5671=5		5.283 55=5		1.320 89=5	
1.623 11=6		0.202 888=6		12.6805=6		6.340 26=6		1.585 07=6	
1.893 63=7		0.236 703=7		14.7939=7		7.396 97=7		1.849 24=7	
2.164 14=8		0.270 518=8		16.9074=8		8.453 68=8		2.113 42=8	
2.434 66=9		0.304 333=9		19.0208=9		9.510 39=9		2.377 60=9	

5. CAPACITY—DRY MEASURE

U. S. dry quarts (qt)	Litres (litre)	U. S. pecks (pk)	Litres (litre)	U. S. pecks (pk)	Dekalitres (dkl)	U. S. bushels (bu)	Hectolitres (hl)	U. S. bushels per acre	Hectolitres per hectare
1=1.1012		1= 8.810		1=0.8810		1=0.352 38		1=0.8708	
2=2.2024		2=17.619		2=1.7619		2=0.704 77		2=1.7415	
3=3.3036		3=26.429		3=2.6429		3=1.057 15		3=2.6123	
4=4.4048		4=35.238		4=3.5238		4=1.409 53		4=3.4830	
5=5.5060		5=44.048		5=4.4048		5=1.761 92		5=4.3538	
6=6.6072		6=52.857		6=5.2857		6=2.114 30		6=5.2245	
7=7.7084		7=61.667		7=6.1667		7=2.466 68		7=6.0953	
8=8.8096		8=70.477		8=7.0477		8=2.819 07		8=6.9660	
9=9.9108		9=79.286		9=7.9286		9=3.171 45		9=7.8368	
0.9081=1		0.113 51=1		1.1351=1		2.8378=1		1.1484=1	
1.8162=2		0.227 03=2		2.2703=2		5.6756=2		2.2969=2	
2.7243=3		0.340 54=3		3.4054=3		8.5135=3		3.4453=3	
3.6324=4		0.454 05=4		4.5405=4		11.3513=4		4.5937=4	
4.5405=5		0.567 56=5		5.6756=5		14.1891=5		5.7421=5	
5.4486=6		0.681 08=6		6.8108=6		17.0269=6		6.8906=6	
6.3567=7		0.794 59=7		7.9459=7		19.8647=7		8.0390=7	
7.2648=8		0.908 10=8		9.0810=8		22.7026=8		9.1874=8	
8.1729=9		1.021 61=9		10.2161=9		25.5404=9		10.3359=9	

6. MASS

Grains (grain)	Grams (g)	Apothecaries' drams (dr ap or ʒ)	Grams (g)	Troy ounces (oz t)	Grams (g)	Avoirdupois ounces (oz avdp)	Grams (g)	Avoirdupois pounds (lb avdp)	Kilograms (kg)
1=0.064 799		1= 3.8879		1= 31.103		1= 28.350		1=0.453 59	
2=0.129 598		2= 7.7759		2= 62.207		2= 56.699		2=0.907 18	
3=0.194 397		3=11.6638		3= 93.310		3= 85.049		3=1.360 78	
4=0.259 196		4=15.5517		4=124.414		4=113.398		4=1.814 37	
5=0.323 995		5=19.4397		5=155.517		5=141.748		5=2.267 96	
6=0.388 794		6=23.3276		6=186.621		6=170.097		6=2.721 55	
7=0.453 592		7=27.2155		7=217.724		7=198.447		7=3.175 15	
8=0.518 391		8=31.1035		8=248.828		8=225.796		8=3.628 74	
9=0.583 190		9=34.9914		9=279.931		9=255.146		9=4.082 33	
15.4324=1		0.257 21=1		0.032 151=1		0.035 274=1		2.204 62=1	
30.8647=2		0.514 41=2		0.064 301=2		0.070 548=2		4.409 24=2	
46.2971=3		0.771 62=3		0.096 452=3		0.105 822=3		6.613 87=3	
61.7294=4		1.028 82=4		0.128 603=4		0.141 096=4		8.818 49=4	
77.1618=5		1.286 03=5		0.160 754=5		0.176 370=5		11.023 11=5	
92.5941=6		1.543 24=6		0.192 904=6		0.211 644=6		13.227 73=6	
108.0265=7		1.800 44=7		0.225 055=7		0.246 918=7		15.432 36=7	
123.4589=8		2.057 65=8		0.257 206=8		0.282 192=8		17.636 98=8	
138.8912=9		2.314 85=9		0.289 357=9		0.317 466=9		19.841 60=9	

COMPARISON OF THE VARIOUS TONS AND POUNDS IN USE IN THE UNITED STATES (FROM 1 TO 9 UNITS)

Troy pounds	Avoirdupois pounds	Kilograms	Short tons	Long tons	Metric tons
1	0.822 857	0.373 24	0.000 411 43	0.000 367 35	0.000 373 24
2	1.645 71	0.746 48	0.000 822 86	0.000 734 69	0.000 746 48
3	2.468 57	1.119 73	0.001 234 29	0.001 102 04	0.001 119 73
4	3.291 43	1.492 97	0.001 645 71	0.001 469 39	0.001 492 97
5	4.114 29	1.866 21	0.002 057 14	0.001 836 73	0.001 866 21
6	4.937 14	2.239 45	0.002 468 57	0.002 204 08	0.002 239 45
7	5.760 00	2.612 69	0.002 880 00	0.002 571 43	0.002 612 69
8	6.582 86	2.985 93	0.003 291 43	0.002 938 78	0.002 985 93
9	7.405 71	3.359 18	0.003 702 86	0.003 306 12	0.003 359 18
1.215 28	1	0.453 59	0.0005	0.000 446 43	0.000 453 59
2.430 56	2	0.907 18	0.0010	0.000 892 86	0.000 907 18
3.645 83	3	1.360 78	0.0015	0.001 339 29	0.001 360 78
4.861 11	4	1.814 37	0.0020	0.001 785 71	0.001 814 37
6.076 39	5	2.267 96	0.0025	0.002 232 14	0.002 267 96
7.291 67	6	2.721 55	0.0030	0.002 678 57	0.002 721 55
8.506 94	7	3.175 15	0.0035	0.003 125 00	0.003 175 15
9.722 22	8	3.628 74	0.0040	0.003 571 43	0.003 628 74
10.937 50	9	4.082 33	0.0045	0.004 017 86	0.004 082 33
2.679 23	2.204 62	1	0.001 102 31	0.000 984 21	0.001
5.358 46	4.409 24	2	0.002 204 62	0.001 968 41	0.002
8.037 69	6.613 87	3	0.003 306 93	0.002 952 62	0.003
10.716 91	8.818 49	4	0.004 409 24	0.003 936 83	0.004
13.396 14	11.023 11	5	0.005 511 56	0.004 921 03	0.005
16.075 37	13.227 73	6	0.006 613 87	0.005 905 24	0.006
18.754 60	15.432 36	7	0.007 716 18	0.006 889 44	0.007
21.433 83	17.636 98	8	0.008 818 49	0.007 873 65	0.008
24.113 06	19.841 60	9	0.009 920 80	0.008 857 86	0.009

Troy pounds	Avoirdupois pounds	Kilograms	Short tons	Long tons	Metric tons
2430.56	2000	907.18	1	0.892 86	0.907 18
4861.11	4000	1814.37	2	1.785 71	1.814 37
7291.67	6000	2721.55	3	2.678 57	2.721 55
9722.22	8000	3628.74	4	3.571 43	3.628 74
12 152.78	10 000	4535.92	5	4.464 29	4.535 92
14 583.33	12 000	5443.11	6	5.357 14	5.443 11
17 013.89	14 000	6350.29	7	6.250 00	6.350 29
19 444.44	16 000	7257.48	8	7.142 86	7.257 48
21 875.00	18 000	8164.66	9	8.035 71	8.164 66
2722.22	2240	1016.05	1.12	1	1.016 05
5444.44	4480	2032.09	2.24	2	2.032 09
8166.67	6720	3048.14	3.36	3	3.048 14
10 888.89	8960	4064.19	4.48	4	4.064 19
13 611.11	11 200	5080.24	5.60	5	5.080 24
16 333.33	13 440	6096.28	6.72	6	6.096 28
19 055.56	15 680	7112.32	7.84	7	7.112 32
21 777.78	17 920	8128.38	8.96	8	8.128 38
24 500.00	20 160	9144.42	10.08	9	9.144 42
2679.23	2204.62	1000	1.102 31	0.984 21	1
5358.46	4409.24	2000	2.204 62	1.968 41	2
8037.69	6613.87	3000	3.306 93	2.952 62	3
10 716.91	8818.49	4000	4.409 24	3.936 83	4
13 396.14	11 023.11	5000	5.511 56	4.921 03	5
16 075.37	13 227.73	6000	6.613 87	5.905 24	6
18 754.60	15 432.36	7000	7.716 18	6.889 44	7
21 433.83	17 636.98	8000	8.818 49	7.873 65	8
24 113.06	19 841.60	9000	9.920 80	8.857 86	9

SPECIAL TABLES

LENGTH—INCHES AND MILLIMETRES—EQUIVALENTS OF DECIMAL AND BINARY FRACTIONS OF AN INCH IN MILLIMETRES

From 1/64 to 1 Inch

½'s	¼'s	8ths	16ths	32ds	64ths	Milli-metres	Decimals of an inch
					1	= 0.397	0.015625
				1	2	= .794	.03125
					3	= 1.191	.046875
			1	2	4	= 1.588	.0625
					5	= 1.984	.078125
				3	6	= 2.381	.09375
					7	= 2.778	.109375
		1	2	4	8	= 3.175	.1250
					9	= 3.572	.140625
				5	10	= 3.969	.15625
					11	= 4.366	.171875
			3	6	12	= 4.763	.1875
					13	= 5.159	.203125
				7	14	= 5.556	.21875
					15	= 5.953	.234375
	1	2	4	8	16	= 6.350	.2500
					17	= 6.747	.265625
				9	18	= 7.144	.28125
					19	= 7.541	.296875
			5	10	20	= 7.938	.3125
					21	= 8.334	.328125
				11	22	= 8.731	.34375
					23	= 9.128	.359375
		3	6	12	24	= 9.525	.3750
					25	= 9.922	.390625
				13	26	= 10.319	.40625
					27	= 10.716	.421875
			7	14	28	= 11.113	.4375
					29	= 11.509	.453125
				15	30	= 11.906	.46875
					31	= 12.303	.484375
1	2	4	8	16	32	= 12.700	.5

Inch	½'s	¼'s	8ths	16ths	32ds	64ths	Milli-metres	Decimals of an inch
						33	= 13.097	0.515625
					17	34	= 13.494	.53125
						35	= 13.891	.546875
				9	18	36	= 14.288	.5625
						37	= 14.684	.578125
					19	38	= 15.081	.59375
						39	= 15.478	.609375
			5	10	20	40	= 15.875	.625
						41	= 16.272	.640625
					21	42	= 16.669	.65625
						43	= 17.066	.671875
				11	22	44	= 17.463	.6875
						45	= 17.859	.703125
					23	46	= 18.256	.71875
						47	= 18.653	.734375
		3	6	12	24	48	= 19.050	.75
						49	= 19.447	.765625
					25	50	= 19.844	.78125
						51	= 20.241	.796875
				13	26	52	= 20.638	.8125
						53	= 21.034	.828125
					27	54	= 21.431	.84375
						55	= 21.828	.859375
			7	14	28	56	= 22.225	.875
						57	= 22.622	.890625
					29	58	= 23.019	.90625
						59	= 23.416	.921875
				15	30	60	= 23.813	.9375
						61	= 24.209	.953125
					31	62	= 24.606	.96875
						63	= 25.003	.984375
1	2	4	8	16	32	64	= 25.400	1.000

SPECIAL TABLES

LENGTH—HUNDREDTHS OF AN INCH TO MILLIMETRES

From 1 to 99 Hundredths

Hundredths of an inch	0	1	2	3	4	5	6	7	8	9
	0	0.254	0.508	0.762	1.016	1.270	1.524	1.778	2.032	2.286
10	2.540	2.794	3.048	3.302	3.556	3.810	4.064	4.318	4.572	4.826
20	5.080	5.334	5.588	5.842	6.096	6.350	6.604	6.858	7.112	7.366
30	7.620	7.874	8.128	8.382	8.636	8.890	9.144	9.398	9.652	9.906
40	10.160	10.414	10.668	10.922	11.176	11.430	11.684	11.938	12.192	12.446
50	12.700	12.954	13.208	13.462	13.716	13.970	14.224	14.478	14.732	14.986
60	15.240	15.494	15.748	16.002	16.256	16.510	16.764	17.018	17.272	17.526
70	17.780	18.034	18.288	18.542	18.796	19.050	19.304	19.558	19.812	20.066
80	20.320	20.574	20.828	21.082	21.336	21.590	21.844	22.098	22.352	22.606
90	22.860	23.114	23.368	23.622	23.876	24.130	24.384	24.638	24.892	25.146

LENGTH—MILLIMETRES TO DECIMALS OF AN INCH

From 1 to 99 Units

Milli- metres	0	1	2	3	4	5	6	7	8	9
	0	0.03937	0.07874	0.11811	0.15748	0.19685	0.23622	0.27559	0.31496	0.35433
10	0.39370	.43307	.47244	.51181	.55118	.59055	.62992	.66929	.70866	.74803
20	.78740	.82677	.86614	.90551	.94488	.98425	1.02362	1.06299	1.10236	1.14173
30	1.18110	1.22047	1.25984	1.29921	1.33858	1.37795	1.41732	1.45669	1.49606	1.53543
40	1.57480	1.61417	1.65354	1.69291	1.73228	1.77165	1.81102	1.85039	1.88976	1.92913
50	1.96850	2.00787	2.04724	2.08661	2.12598	2.16535	2.20472	2.24409	2.28346	2.32283
60	2.36220	2.40157	2.44094	2.48031	2.51968	2.55905	2.59842	2.63779	2.67716	2.71653
70	2.75590	2.79527	2.83464	2.87401	2.91338	2.95275	2.99212	3.03149	3.07086	3.11023
80	3.14960	3.18897	3.22834	3.26771	3.30708	3.34645	3.38582	3.42519	3.45456	3.50393
90	3.54330	3.58267	3.62204	3.66141	3.70078	3.74015	3.77952	3.81889	3.85826	3.89763

LENGTH—UNITED STATES NAUTICAL MILES, INTERNATIONAL NAUTICAL MILES, AND KILOMETRES

Basic relations 1 U.S. nautical mile = 1.853 248 kilometres. 1 U.S. nautical mile = 1.000 673 9 int. nautical miles. 1 International nautical mile = 1.852 kilometres.

U.S. nautical miles	Int. nautical miles	Kilometres	U.S. nautical miles	Int. nautical miles	Kilometres	U.S. nautical miles	Int. nautical miles	Kilometres
0				0				0
1	1.0007	1.8532	0.9993	1	1.8520	0.5396	0.5400	1
2	2.0013	3.7065	1.9987	2	3.7040	1.0792	1.0799	2
3	3.0020	5.5597	2.9980	3	5.5560	1.6188	1.6199	3
4	4.0027	7.4130	3.9973	4	7.4080	2.1584	2.1598	4
5	5.0034	9.2662	4.9966	5	9.2600	2.6980	2.6998	5
6	6.0040	11.1195	5.9960	6	11.1120	3.2376	3.2397	6
7	7.0047	12.9727	6.9953	7	12.9640	3.7771	3.7797	7
8	8.0054	14.8260	7.9946	8	14.8160	4.3167	4.3197	8
9	9.0061	16.6792	8.9939	9	16.6680	4.8563	4.8596	9
10	10.0067	18.5325	9.9933	10	18.5200	5.3959	5.3996	10
11	11.0074	20.3857	10.9926	11	20.3720	5.9355	5.9395	11
12	12.0081	22.2390	11.9919	12	22.2240	6.4751	6.4795	12
13	13.0088	24.0922	12.9912	13	24.0760	7.0147	7.0194	13
14	14.0094	25.9455	13.9906	14	25.9280	7.5543	7.5594	14
15	15.0101	27.7987	14.9899	15	27.7800	8.0939	8.0994	15
16	16.0108	29.6520	15.9892	16	29.6320	8.6335	8.6393	16
17	17.0115	31.5052	16.9886	17	31.4840	9.1731	9.1793	17
18	18.0121	33.3585	17.9879	18	33.3360	9.7127	9.7192	18
19	19.0128	35.2117	18.9872	19	35.1880	10.2523	10.2592	19
20	20.0135	37.0650	19.9865	20	37.0400	10.7919	10.7991	20
21	21.0142	38.9182	20.9859	21	38.8920	11.3315	11.3391	21
22	22.0148	40.7715	21.9852	22	40.7440	11.8711	11.8790	22
23	23.0155	42.6247	22.9845	23	42.5960	12.4106	12.4190	23
24	24.0162	44.4780	23.9838	24	44.4480	12.9502	12.9590	24
25	25.0168	46.3312	24.9832	25	46.3000	13.4898	13.4989	25
26	26.0175	48.1844	25.9825	26	48.1520	14.0294	14.0389	26
27	27.0182	50.0377	26.9818	27	50.0040	14.5690	14.5788	27
28	28.0189	51.8909	27.9811	28	51.8560	15.1086	15.1188	28
29	29.0195	53.7442	28.9805	29	53.7080	15.6482	15.6587	29
30	30.0202	55.5974	29.9798	30	55.5600	16.1878	16.1987	30
31	31.0209	57.4507	30.9791	31	57.4120	16.7274	16.7387	31
32	32.0216	59.3039	31.9785	32	59.2640	17.2670	17.2786	32
33	33.0222	61.1572	32.9778	33	61.1160	17.8066	17.8186	33
34	34.0229	63.0104	33.9771	34	62.9680	18.3462	18.3585	34
35	35.0236	64.8637	34.9764	35	64.8200	18.8858	18.8985	35
36	36.0243	66.7169	35.9758	36	66.6720	19.4254	19.4384	36
37	37.0249	68.5702	36.9751	37	68.5240	19.9649	19.9784	37
38	38.0256	70.4234	37.9744	38	70.3760	20.5045	20.5184	38
39	39.0263	72.2767	38.9737	39	72.2280	21.0441	21.0583	39
40	40.0270	74.1299	39.9731	40	74.0800	21.5837	21.5983	40
41	41.0276	75.9832	40.9724	41	75.9320	22.1233	22.1382	41
42	42.0283	77.8364	41.9717	42	77.7840	22.6629	22.6782	42
43	43.0290	79.6897	42.9710	43	79.6360	23.2025	23.2181	43
44	44.0297	81.5429	43.9704	44	81.4880	23.7421	23.7581	44
45	45.0303	83.3962	44.9697	45	83.3400	24.2817	24.2981	45
46	46.0310	85.2494	45.9690	46	85.1920	24.8213	24.8380	46
47	47.0317	87.1027	46.9683	47	87.0440	25.3609	25.3780	47
48	48.0323	88.9559	47.9677	48	88.8960	25.9005	25.9179	48
49	49.0330	90.8092	48.9670	49	90.7480	26.4401	26.4579	49

No.						
50	26.9978	26.9797	92.6000	49.9663	92.6624	50.0337
51	27.5378	27.5193	94.4520	50.9657	94.5156	51.0344
52	28.0778	28.0588	96.3040	51.9650	96.3689	52.0350
53	28.6177	28.5984	98.1560	52.9643	98.2221	53.0357
54	29.1577	29.1380	100.0080	53.9636	100.0754	54.0364
55	29.6976	29.6776	101.8600	54.9630	101.9286	55.0371
56	30.2376	30.2172	103.7120	55.9623	103.7819	56.0377
57	30.7775	30.7568	105.5640	56.9616	105.6351	57.0384
58	31.3175	31.2964	107.4160	57.9609	107.4884	58.0391
59	31.8575	31.8360	109.2680	58.9603	109.3416	59.0398
60	32.3974	32.3756	111.1200	59.9596	111.1949	60.0404
61	32.9374	32.9152	112.9720	60.9589	113.0481	61.0411
62	33.4773	33.4548	114.8240	61.9582	114.9014	62.0418
63	34.0173	33.9944	116.6760	62.9576	116.7546	63.0425
64	34.5572	34.5340	118.5280	63.9569	118.6079	64.0431
65	35.0972	35.0736	120.3800	64.9562	120.4611	65.0438
66	35.6371	35.6132	122.2320	65.9556	122.3144	66.0445
67	36.1771	36.1527	124.0840	66.9549	124.1676	67.0452
68	36.7171	36.6923	125.9360	67.9542	126.0209	68.0458
69	37.2570	37.2319	127.7880	68.9535	127.8741	69.0465
70	37.7970	37.7715	129.6400	69.9529	129.7274	70.0472
71	38.3369	38.3111	131.4920	70.9522	131.5806	71.0478
72	38.8769	38.8507	133.3440	71.9515	133.4339	72.0485
73	39.4168	39.3903	135.1960	72.9508	135.2871	73.0492
74	39.9568	39.9299	137.0480	73.9502	137.1404	74.0499
75	40.4968	40.4695	138.9000	74.9495	138.9936	75.0505
76	41.0367	41.0091	140.7520	75.9488	140.8468	76.0512
77	41.5767	41.5487	142.6040	76.9481	142.7001	77.0519
78	42.1166	42.0883	144.4560	77.9475	144.5533	78.0526
79	42.6566	42.6279	146.3080	78.9468	146.4066	79.0532
80	43.1965	43.1675	148.1600	79.9461	148.2598	80.0539
81	43.7365	43.7070	150.0120	80.9455	150.1131	81.0546
82	44.2764	44.2466	151.8640	81.9448	151.9663	82.0553
83	44.8164	44.7862	153.7160	82.9441	153.8196	83.0559
84	45.3564	45.3258	155.5680	83.9434	155.6728	84.0566
85	45.8963	45.8654	157.4200	84.9428	157.5261	85.0573
86	46.4363	46.4050	159.2720	85.9421	159.3793	86.0580
87	46.9762	46.9446	161.1240	86.9414	161.2326	87.0586
88	47.5162	47.4842	162.9760	87.9407	163.0858	88.0593
89	48.0562	48.0238	164.8280	88.9401	164.9391	89.0600
90	48.5961	48.5634	166.6800	89.9394	166.7923	90.0607
91	49.1361	49.1030	168.5320	90.9387	168.6456	91.0613
92	49.6760	49.6426	170.3840	91.9380	170.4988	92.0620
93	50.2160	50.1822	172.2360	92.9374	172.3521	93.0627
94	50.7559	50.7218	174.0880	93.9367	174.2053	94.0633
95	51.2959	51.2614	175.9400	94.9360	176.0586	95.0640
96	51.8359	51.8009	177.7920	95.9354	177.9118	96.0647
97	52.3758	52.3405	179.6440	96.9347	179.7651	97.0654
98	52.9158	52.8801	181.4960	97.9340	181.6183	98.0660
99	53.4557	53.4197	183.3480	98.9333	183.4716	99.0667
100	53.9957	53.9593	185.2000	99.9327	185.3248	100.0674

LENGTH—MILLIMETRES TO INCHES

[From 0.00 to 25.40 millimetres by 0.01 millimetre. 1 millimetre = 0.03937 inch.]

Milli-metres	Hundredths of millimetres				
	0.00	0.01	0.02	0.03	0.04
	Inches	Inches	Inches	Inches	Inches
0.00	0. 000000	0. 000394	0. 000787	0. 001181	0. 001575
0. 10	. 003937	. 004331	. 004724	. 005118	. 005512
0. 20	. 007874	. 008268	. 008661	. 009055	. 009449
0. 30	. 011811	. 012205	. 012598	. 012992	. 013386
0. 40	. 015748	. 016142	. 016535	. 016929	. 017323
0.50	0. 019685	0. 020079	0. 020472	0. 020866	0. 021260
0. 60	. 023622	. 024016	. 024409	. 024803	. 025197
0. 70	. 027559	. 027953	. 028346	. 028740	. 029134
0. 80	. 031496	. 031890	. 032283	. 032677	. 033071
0. 90	. 035433	. 035827	. 036220	. 036614	. 037008
1.00	0. 03937	0. 03976	0. 04016	0. 04055	0. 04094
1. 10	. 04331	. 04370	. 04409	. 04449	. 04488
1. 20	. 04724	. 04764	. 04803	. 04843	. 04882
1. 30	. 05118	. 05157	. 05197	. 05236	. 05276
1. 40	. 05512	. 05551	. 05591	. 05630	. 05669
1.50	0. 05906	0. 05945	0. 05984	0. 06024	0. 06063
1. 60	. 06299	. 06339	. 06378	. 06417	. 06457
1. 70	. 06693	. 06732	. 06772	. 06811	. 06850
1. 80	. 07087	. 07126	. 07165	. 07205	. 07244
1. 90	. 07480	. 07520	. 07559	. 07598	. 07638
2.00	0. 07874	0. 07913	0. 07953	0. 07992	0. 08031
2. 10	. 08268	. 08307	. 08346	. 08386	. 08425
2. 20	. 08661	. 08701	. 08740	. 08780	. 08819
2. 30	. 09055	. 09094	. 09134	. 09173	. 09213
2. 40	. 09449	. 09488	. 09528	. 09567	. 09606
2.50	0. 09842	0. 09882	0. 09921	0. 09961	0. 10000
2. 60	. 10236	. 10276	. 10315	. 10354	. 10394
2. 70	. 10630	. 10669	. 10709	. 10748	. 10787
2. 80	. 11024	. 11063	. 11102	. 11142	. 11181
2. 90	. 11417	. 11457	. 11496	. 11535	. 11575
3.00	0. 11811	0. 11850	0. 11890	0. 11929	0. 11968
3. 10	. 12205	. 12244	. 12283	. 12323	. 12362
3. 20	. 12598	. 12638	. 12677	. 12717	. 12756
3. 30	. 12992	. 13031	. 13071	. 13110	. 13150
3. 40	. 13386	. 13425	. 13465	. 13504	. 13543

LENGTH—MILLIMETRES TO INCHES—continued

[From 0.00 to 25.40 millimetres by 0.01 millimetre. 1 millimetre = 0.03937 inch.]

Milli-metres	Hundredths of millimetres				
	0.05	0.06	0.07	0.08	0.09
	Inches	Inches	Inches	Inches	Inches
0.00	0.001968	0.002362	0.002756	0.003150	0.003543
0.10	.005906	.006299	.006693	.007087	.007480
0.20	.009842	.010236	.010630	.011024	.011417
0.30	.013780	.014173	.014567	.014961	.015354
0.40	.017716	.018110	.018504	.018898	.019291
0.50	0.021654	0.022047	0.022441	0.022835	0.023228
0.60	.025590	.025984	.026378	.026772	.027165
0.70	.029528	.029921	.030315	.030709	.031102
0.80	.033464	.033858	.034252	.034646	.035039
0.90	.037402	.037795	.038189	.038583	.038976
1.00	0.04134	0.04173	0.04213	0.04252	0.04291
1.10	.04528	.04567	.04606	.04646	.04685
1.20	.04921	.04961	.05000	.05039	.05079
1.30	.05315	.05354	.05394	.05433	.05472
1.40	.05709	.05748	.05787	.05827	.05866
1.50	0.06102	0.06142	0.06181	0.06220	0.06260
1.60	.06496	.06535	.06575	.06614	.06654
1.70	.06890	.06929	.06968	.07008	.07047
1.80	.07283	.07323	.07362	.07402	.07441
1.90	.07677	.07717	.07756	.07795	.07835
2.00	0.08071	0.08110	0.08150	0.08189	0.08228
2.10	.08465	.08504	.08543	.08583	.08622
2.20	.08858	.08898	.08937	.08976	.09016
2.30	.09252	.09291	.09331	.09370	.09409
2.40	.09646	.09685	.09724	.09764	.09803
2.50	0.10039	0.10079	0.10118	0.10157	0.10197
2.60	.10433	.10472	.10512	.10551	.10591
2.70	.10827	.10866	.10905	.10945	.10984
2.80	.11220	.11260	.11299	.11339	.11378
2.90	.11614	.11654	.11693	.11732	.11772
3.00	0.12008	0.12047	0.12087	0.12126	0.12165
3.10	.12402	.12441	.12480	.12520	.12559
3.20	.12795	.12835	.12874	.12913	.12953
3.30	.13189	.13228	.13268	.13307	.13346
3.40	.13583	.13622	.13661	.13701	.13740

LENGTH—MILLIMETRES TO INCHES—continued

[From 0.00 to 25.40 millimetres by 0.01 millimetre. 1 millimetre = 0.03937 inch.]

Milli-metres	Hundredths of millimetres				
	0.00	0.01	0.02	0.03	0.04
	Inches	Inches	Inches	Inches	Inches
3.50	0. 13780	0. 13819	0. 13858	0. 13898	0.13937
3. 60	. 14173	. 14213	. 14252	. 14291	. 14331
3. 70	. 14567	. 14606	. 14646	. 14685	. 14724
3. 80	. 14961	. 15000	. 15039	. 15079	. 15118
3. 90	. 15354	. 15394	. 15433	. 15472	. 15512
4.00	0. 15748	0. 15787	0. 15827	0. 15866	0. 15905
4. 10	. 16142	. 16181	. 16220	. 16260	. 16299
4. 20	. 16535	. 16575	. 16614	. 16654	. 16693
4. 30	. 16929	. 16968	. 17008	. 17047	. 17087
4. 40	. 17323	. 17362	. 17402	. 17441	. 17480
4.50	0. 17716	0. 17756	0. 17795	0. 17835	0. 17874
4. 60	. 18110	. 18150	. 18189	. 18228	. 18268
4. 70	. 18504	. 18543	. 18583	. 18622	. 18661
4. 80	. 18898	. 18937	. 18976	. 19016	. 19055
4. 90	. 19291	. 19331	. 19370	. 19409	. 19449
5.00	0. 19685	0. 19724	0. 19764	0. 19803	0. 19842
5. 10	. 20079	. 20118	. 20157	. 20197	. 20236
5. 20	. 20472	. 20512	. 20551	. 20591	. 20630
5. 30	. 20866	. 20905	. 20945	. 20984	. 21024
5. 40	. 21260	. 21299	. 21339	. 21378	. 21417
5.50	0. 21654	0. 21693	0. 21732	0. 21772	0.21811
5. 60	. 22047	. 22087	. 22126	. 22165	. 22205
5. 70	. 22441	. 22480	. 22520	. 22559	. 22598
5. 80	. 22835	. 22874	. 22913	. 22953	. 22992
5. 90	. 23228	. 23268	. 23307	. 23346	. 23386
6.00	0. 23622	0. 23661	0. 23701	0. 23740	0. 23779
6. 10	. 24016	. 24055	. 24094	. 24134	. 24173
6. 20	. 24409	. 24449	. 24488	. 24528	. 24567
6. 30	. 24803	. 24842	. 24882	. 24921	. 24961
6. 40	. 25197	. 25236	. 25276	. 25315	. 25354
6.50	0. 25590	0. 25630	0. 25669	0. 25709	0. 25748
6. 60	. 25984	. 26024	. 26063	. 26102	. 26142
6. 70	. 26378	. 26417	. 26457	. 26496	. 26535
6. 80	. 26772	. 26811	. 26850	. 26890	. 26929
6. 90	. 27165	. 27205	. 27244	. 27283	. 27323

LENGTH—MILLIMETRES TO INCHES—continued

[From 0.00 to 25.40 millimetres by 0.01 millimetre. 1 millimetre = 0.03937 inch.]

Milli- metres	Hundredths of millimetres				
	0.05	**0.06**	**0.07**	**0.08**	**0.09**
	Inches	Inches	Inches	Inches	Inches
3.50	0. 13976	0. 14016	0. 14055	0. 14094	0. 14134
3. 60	. 14370	. 14409	. 14449	. 14488	. 14528
3. 70	. 14764	. 14803	. 14842	. 14882	. 14921
3. 80	. 15157	. 15197	. 15236	. 15276	. 15315
3. 90	. 15551	. 15591	. 15630	. 15669	. 15709
4.00	0. 15945	0. 15984	0. 16024	0. 16063	0. 16102
4. 10	. 16339	. 16378	. 16417	. 16457	. 16496
4. 20	. 16732	. 16772	. 16811	. 16850	. 16890
4. 30	. 17126	. 17165	. 17205	. 17244	. 17283
4. 40	. 17520	. 17559	. 17598	. 17638	. 17677
4.50	0. 17913	0. 17953	0. 17992	0. 18031	0. 18071
4. 60	. 18307	. 18346	. 18386	. 18425	. 18465
4. 70	. 18701	. 18740	. 18779	. 18819	. 18858
4. 80	. 19094	. 19134	. 19173	. 19213	. 19252
4. 90	. 19488	. 19528	. 19567	. 19606	. 19646
5.00	0. 19882	0. 19921	0. 19961	0. 20000	0. 20039
5. 10	. 20276	. 20315	. 20354	. 20394	. 20433
5. 20	. 20669	. 20709	. 20748	. 20787	. 20827
5. 30	. 21063	. 21102	. 21142	. 21181	. 21220
5. 40	. 21457	. 21496	. 21535	. 21575	. 21614
5.50	0. 21850	0. 21890	0. 21929	0. 21968	0. 22008
5. 60	. 22244	. 22283	. 22323	. 22362	. 22402
5. 70	. 22638	. 22677	. 22716	. 22756	. 22795
5. 80	. 23031	. 23071	. 23110	. 23150	. 23189
5. 90	. 23425	. 23465	. 23504	. 23543	. 23583
6.00	0. 23819	0. 23858	0. 23898	0. 23937	0. 23976
6. 10	. 24213	. 24252	. 24291	. 24331	. 24370
6. 20	. 24606	. 24646	. 24685	. 24724	. 24764
6. 30	. 25000	. 25039	. 25079	. 25118	. 25157
6. 40	. 25394	. 25433	. 25472	. 25512	. 25551
6.50	0. 25787	0. 25827	0. 25866	0. 25905	0. 25945
6. 60	. 26181	. 26220	. 26260	. 26299	. 26339
6. 70	. 26575	. 26614	. 26653	. 26693	. 26732
6. 80	. 26968	. 27008	. 27047	. 27087	. 27126
6. 90	. 27362	. 27402	. 27441	. 27480	. 27520

LENGTH—MILLIMETRES TO INCHES—continued

[From 0.00 to 25.40 millimetres by 0.01 millimetre. 1 millimetre = 0.03937 inch.]

Milli-metres	Hundredths of millimetres				
	0.00	0.01	0.02	0.03	0.04
	Inches	Inches	Inches	Inches	Inches
7.00	0. 27559	0. 27598	0. 27638	0. 27677	0. 27716
7. 10	. 27953	. 27992	. 28031	. 28071	. 28110
7. 20	. 28346	. 28386	. 28425	. 28465	. 28504
7. 30	. 28740	. 28779	. 28819	. 28858	. 28898
7. 40	. 29134	. 29173	. 29213	. 29252	. 29291
7.50	0. 29528	0. 29567	0. 29606	0. 29646	0. 29685
7. 60	. 29921	. 29961	. 30000	. 30039	. 30079
7. 70	. 30315	. 30354	. 30394	. 30433	. 30472
7. 80	. 30709	. 30748	. 30787	. 30827	. 30866
7. 90	. 31102	. 31142	. 31181	. 31220	. 31260
8.00	0. 31496	0. 31535	0. 31575	0. 31614	0. 31653
8. 10	. 31890	. 31929	. 31968	. 32008	. 32047
8. 20	. 32283	. 32323	. 32362	. 32402	. 32441
8. 30	. 32677	. 32716	. 32756	. 32795	. 32835
8. 40	. 33071	. 33110	. 33150	. 33189	. 33228
8.50	0. 33464	0. 33504	0. 33543	0. 33583	0. 33622
8. 60	. 33858	. 33898	. 33937	. 33976	. 34016
8. 70	. 34252	. 34291	. 34331	. 34370	. 34409
8. 80	. 34646	. 34685	. 34724	. 34764	. 34803
8. 90	. 35039	. 35079	. 35118	. 35157	. 35197
9.00	0. 35433	0. 35472	0. 35512	0. 35551	0. 35590
9. 10	. 35827	. 35866	. 35905	. 35945	. 35984
9. 20	. 36220	. 36260	. 36299	. 36339	. 36378
9. 30	. 36614	. 36653	. 36693	. 36732	. 36772
9. 40	. 37008	. 37047	. 37087	. 37126	. 37165
9.50	0. 37402	0. 37441	0. 37480	0. 37520	0. 37559
9. 60	. 37795	. 37835	. 37874	. 37913	. 37953
9. 70	. 38189	. 38228	. 38268	. 38307	. 38346
9. 80	. 38583	. 38622	. 38661	. 38701	. 38740
9. 90	. 38976	. 39016	. 39055	. 39094	. 39134
10.00	0. 39370	0. 39409	0. 39449	0. 39488	0. 39527
10. 10	. 39764	. 39803	. 39842	. 39882	. 39921
10. 20	. 40157	. 40197	. 40236	. 40276	. 40315
10. 30	. 40551	. 40590	. 40630	. 40669	. 40709
10. 40	. 40945	. 40984	. 41024	. 41063	. 41102

LENGTH—MILLIMETRES TO INCHES—continued

[From 0.00 to 25.40 millimetres by 0.01 millimetre. 1 millimetre = 0.03937 inch.]

Milli-metres	Hundredths of millimetres				
	0.05	0.06	0.07	0.08	0.09
	Inches	Inches	Inches	Inches	Inches
7.00	0. 27756	0. 27795	0. 27835	0. 27874	0. 27913
7. 10	. 28150	. 28189	. 28228	. 28268	. 28307
7. 20	. 28543	. 28583	. 28622	. 28661	. 28701
7. 30	. 28937	. 28976	. 29016	. 29055	. 29094
7. 40	. 29331	. 29370	. 29409	. 29449	. 29488
7.50	0. 29724	0. 29764	0. 29803	0. 29842	0. 29882
7. 60	. 30118	. 30157	. 30197	. 30236	. 30276
7. 70	. 30512	. 30551	. 30590	. 30630	. 30669
7. 80	. 30905	. 30945	. 30984	. 31024	. 31063
7. 90	. 31299	. 31339	. 31378	. 31417	. 31457
8.00	0. 31693	0. 31732	0. 31772	0. 31811	0. 31850
8. 10	. 32087	. 32126	. 32165	. 32205	. 32244
8. 20	. 32480	. 32520	. 32559	. 32598	. 32638
8. 30	. 32874	. 32913	. 32953	. 32992	. 33031
8. 40	. 33268	. 33307	. 33346	. 33386	. 33425
8.50	0. 33661	0. 33701	0. 33740	0. 33779	0. 33819
8. 60	. 34055	. 34094	. 34134	. 34173	. 34213
8. 70	. 34449	. 34488	. 34527	. 34567	. 34606
8. 80	. 34842	. 34882	. 34921	. 34961	. 35000
8. 90	. 35236	. 35276	. 35315	. 35354	. 35394
9.00	0. 35630	0. 35669	0. 35709	0. 35748	0. 35787
9. 10	. 36024	. 36063	. 36102	. 36142	. 36181
9. 20	. 36417	. 36457	. 36496	. 36535	. 36575
9. 30	. 36811	. 36850	. 36890	. 36929	. 36968
9. 40	. 37205	. 37244	. 37283	. 37323	. 37362
9.50	0. 37598	0. 37638	0. 37677	0. 37716	0. 37756
9. 60	. 37992	. 38031	. 38071	. 38110	. 38150
9. 70	. 38386	. 38425	. 38464	. 38504	. 38543
9. 80	. 38779	. 38819	. 38858	. 38898	. 38937
9. 90	. 39173	. 39213	. 39252	. 39291	. 39331
10.00	0. 39567	0. 39606	0. 39646	0. 39685	0. 39724
10. 10	. 39961	. 40000	. 40039	. 40079	. 40118
10. 20	. 40354	. 40394	. 40433	. 40472	. 40512
10. 30	. 40748	. 40787	. 40827	. 40866	. 40905
10. 40	. 41142	. 41181	. 41220	. 41260	. 41299

LENGTH—MILLIMETRES TO INCHES—continued

[From 0.00 to 25.40 millimetres by 0.01 millimetre. 1 millimetre = 0.03937 inch.]

Milli-metres	Hundredths of millimetres				
	0.00	0.01	0.02	0.03	0.04
	Inches	Inches	Inches	Inches	Inches
10.50	0. 41338	0. 41378	0. 41417	0. 41457	0. 41496
10. 60	. 41732	. 41772	. 41811	. 41850	. 41890
10. 70	. 42126	. 42165	. 42205	. 42244	. 42283
10. 80	. 42520	. 42559	. 42598	. 42638	. 42677
10. 90	. 42913	. 42953	. 42992	. 43031	. 43071
11.00	0. 43307	0. 43346	0. 43386	0. 43425	0. 43464
11. 10	. 43701	. 43740	. 43779	. 43819	. 43858
11. 20	. 44094	. 44134	. 44173	. 44213	. 44252
11. 30	. 44488	. 44527	. 44567	. 44606	. 44646
11. 40	. 44882	. 44921	. 44961	. 45000	. 45039
11.50	0. 45276	0. 45315	0. 45354	0. 45394	0. 45433
11. 60	. 45669	. 45709	. 45748	. 45787	. 45827
11. 70	. 46063	. 46102	. 46142	. 46181	. 46220
11. 80	. 46457	. 46496	. 46535	. 46575	. 46614
11. 90	. 46850	. 46890	. 46929	. 46968	. 47008
12.00	0. 47244	0. 47283	0. 47323	0. 47362	0. 47401
12. 10	. 47638	. 47677	. 47716	. 47756	. 47795
12. 20	. 48031	. 48071	. 48110	. 48150	. 48189
12. 30	. 48425	. 48464	. 48504	. 48543	. 48583
12. 40	. 48819	. 48858	. 48898	. 48937	. 48976
12.50	0. 49212	0. 49252	0. 49291	0. 49331	0. 49370
12. 60	. 49606	. 49646	. 49685	. 49724	. 49764
12. 70	. 50000	. 50039	. 50079	. 50118	. 50157
12. 80	. 50394	. 50433	. 50472	. 50512	. 50551
12. 90	. 50787	. 50827	. 50866	. 50905	. 50945
13.00	0. 51181	0. 51220	0. 51260	0. 51299	0. 51338
13. 10	. 51575	. 51614	. 51653	. 51693	. 51732
13. 20	. 51968	. 52008	. 52047	. 52087	. 52126
13. 30	. 52362	. 52401	. 52441	. 52480	. 52520
13. 40	. 52756	. 52795	. 52835	. 52874	. 52913
13.50	0. 53150	0. 53189	0. 53228	0. 53268	0. 53307
13. 60	. 53543	. 53583	. 53622	. 53661	. 53701
13. 70	. 53937	. 53976	. 54016	. 54055	. 54094
13. 80	. 54331	. 54370	. 54409	. 54449	. 54488
13. 90	. 54724	. 54764	. 54803	. 54842	. 54882

LENGTH—MILLIMETRES TO INCHES—continued

[From 0.00 to 25.40 millimetres by 0.01 millimetre. 1 millimetre = 0.03937 inch.]

Milli-metres	Hundredths of millimetres				
	0.05	0.06	0.07	0.08	0.09
	Inches	Inches	Inches	Inches	Inches
10.50	0. 41535	0. 41575	0. 41614	0. 41653	0. 41693
10. 60	. 41929	. 41968	. 42008	. 42047	. 42087
10. 70	. 42323	. 42362	. 42401	. 42441	. 42480
10. 80	. 42716	. 42756	. 42795	. 42835	. 42874
10. 90	. 43110	. 43150	. 43189	. 43228	. 43268
11.00	0. 43504	0. 43543	0. 43583	0. 43622	0. 43661
11. 10	. 43898	. 43937	. 43976	. 44016	. 44055
11. 20	. 44291	. 44331	. 44370	. 44409	. 44449
11. 30	. 44685	. 44724	. 44764	. 44803	. 44842
11. 40	. 45079	. 45118	. 45157	. 45197	. 45236
11.50	0. 45472	0. 45512	0. 45551	0. 45590	0. 45630
11. 60	. 45866	. 45905	. 45945	. 45984	. 46024
11. 70	. 46260	. 46299	. 46338	. 46378	. 46417
11. 80	. 46653	. 46693	. 46732	. 46772	. 46811
11. 90	. 47047	. 47087	. 47126	. 47165	. 47205
12.00	0. 47441	0. 47480	0. 47520	0. 47559	0. 47598
12. 10	. 47835	. 47874	. 47913	. 47953	. 47992
12. 20	. 48228	. 48268	. 48307	. 48346	. 48386
12. 30	. 48622	. 48661	. 48701	. 48740	. 48779
12. 40	. 49016	. 49055	. 49094	. 49134	. 49173
12.50	0. 49409	0. 49449	0. 49488	0. 49527	0. 49567
12. 60	. 49803	. 49842	. 49882	. 49921	. 49961
12. 70	. 50197	. 50236	. 50275	. 50315	. 50354
12. 80	. 50590	. 50630	. 50669	. 50709	. 50748
12. 90	. 50984	. 51024	. 51063	. 51102	. 51142
13.00	0. 51378	0. 51417	0. 51457	0. 51496	0. 51535
13. 10	. 51772	. 51811	. 51850	. 51890	. 51929
13. 20	. 52165	. 52205	. 52244	. 52283	. 52323
13. 30	. 52559	. 52598	. 52638	. 52677	. 52716
13. 40	. 52953	. 52992	. 53031	. 53071	. 53110
13.50	0. 53346	0. 53386	0. 53425	0. 53464	0. 53504
13. 60	. 53740	. 53779	. 53819	. 53858	. 53898
13. 70	. 54134	. 54173	. 54212	. 54252	. 54291
13. 80	. 54527	. 54567	. 54606	. 54646	. 54685
13. 90	. 54921	. 54961	. 55000	. 55039	. 55079

LENGTH—MILLIMETRES TO INCHES—continued

[From 0.00 to 25.40 millimetres by 0.01 millimetre. 1 millimetre = 0.03937 inch.]

Milli-metres	Hundredths of millimetres				
	0.00	0.01	0.02	0.03	0.04
	Inches	Inches	Inches	Inches	Inches
14.00	0. 55118	0. 55157	0. 55197	0. 55236	0. 55275
14. 10	. 55512	. 55551	. 55590	. 55630	. 55669
14. 20	. 55905	. 55945	. 55984	. 56024	. 56063
14. 30	. 56299	. 56338	. 56378	. 56417	. 56457
14. 40	. 56693	. 56732	. 56772	. 56811	. 56850
14.50	0. 57086	0. 57126	0. 57165	0. 57205	0. 57244
14. 60	. 57480	. 57520	. 57559	. 57598	. 57638
14. 70	. 57874	. 57913	. 57953	. 57992	. 58031
14. 80	. 58268	. 58307	. 58346	. 58386	. 58425
14. 90	. 58661	. 58701	. 58740	. 58779	. 58819
15.00	0. 59055	0. 59094	0. 59134	0. 59173	0. 59212
15. 10	. 59449	. 59488	. 59527	. 59567	. 59606
15. 20	. 59842	. 59882	. 59921	. 59961	. 60000
15. 30	. 60236	. 60275	. 60315	. 60354	. 60394
15. 40	. 60630	. 60669	. 60709	. 60748	. 60787
15.50	0. 61024	0. 61063	0. 61102	0. 61142	0. 61181
15. 60	. 61417	. 61457	. 61496	. 61535	. 61575
15. 70	. 61811	. 61850	. 61890	. 61929	. 61968
15. 80	. 62205	. 62244	. 62283	. 62323	. 62362
15. 90	. 62598	. 62638	. 62677	. 62716	. 62756
16.00	0. 62992	0. 63031	0. 63071	0. 63110	0. 63149
16. 10	. 63386	. 63425	. 63464	. 63504	. 63543
16. 20	. 63779	. 63819	. 63858	. 63898	. 63937
16. 30	. 64173	. 64212	. 64252	. 64291	. 64331
16. 40	. 64567	. 64606	. 64646	. 64685	. 64724
16.50	0. 64960	0. 65000	0. 65039	0. 65079	0. 65118
16. 60	. 65354	. 65394	. 65433	. 65472	. 65512
16. 70	. 65748	. 65787	. 65827	. 65866	. 65905
16. 80	. 66142	. 66181	. 66220	. 66260	. 66299
16. 90	. 66535	. 66575	. 66614	. 66653	. 66693
17.00	0. 66929	0. 66968	0. 67008	0. 67047	0. 67086
17. 10	. 67323	. 67362	. 67401	. 67441	. 67480
17. 20	. 67716	. 67756	. 67795	. 67835	. 67874
17. 30	. 68110	. 68149	. 68189	. 68228	. 68268
17. 40	. 68504	. 68543	. 68583	. 68622	. 68661

LENGTH—MILLIMETRES TO INCHES—continued

[From 0.00 to 25.40 millimetres by 0.01 millimetre. 1 millimetre = 0.03937 inch.]

Milli-metres	Hundredths of millimetres				
	0.05	0.06	0.07	0.08	0.09
	Inches	Inches	Inches	Inches	Inches
14.00	0. 55315	0. 55354	0. 55394	0. 55433	0. 55472
14. 10	. 55709	. 55748	. 55787	. 55827	. 55866
14. 20	. 56102	. 56142	. 56181	. 56220	. 56260
14. 30	. 56496	. 56535	. 56575	. 56614	. 56653
14. 40	. 56890	. 56929	. 56968	. 57008	. 57047
14.50	0. 57283	0. 57323	0. 57362	0. 57401	. 57441
14. 60	. 57677	. 57716	. 57756	. 57795	. 57835
14. 70	. 58071	. 58110	. 58149	. 58189	. 58228
14. 80	. 58464	. 58504	. 58543	. 58583	. 58622
14. 90	. 58858	. 58898	. 58937	. 58976	. 59016
15.00	0. 59252	0. 59291	0. 59331	0. 59370	0. 59409
15. 10	. 59646	. 59685	. 59724	. 59764	. 59803
15. 20	. 60039	. 60079	. 60118	. 60157	. 60197
15. 30	. 60433	. 60472	. 60512	. 60551	. 60590
15. 40	. 60827	. 60866	. 60905	. 60945	. 60984
15.50	0. 61220	0. 61260	0. 61299	0. 61338	0. 61378
15. 60	. 61614	. 61653	. 61693	. 61732	. 61772
15. 70	. 62008	. 62047	. 62086	. 62126	. 62165
15. 80	. 62401	. 62441	. 62480.	. 62520	. 62559
15. 90	. 62795	. 62835	. 62874	. 62913	. 62953
16.00	0. 63189	0. 63228	0. 63268	0. 63307	0. 63346
16. 10	. 63583	. 63622	. 63661	. 63701	. 63740
16. 20	. 63976	. 64016	. 64055	. 64094	. 64134
16. 30	. 64370	. 64409	. 64449	. 64488	. 64527
16. 40	. 64764	. 64803	. 64842	. 64882	. 64921
16.50	0. 65157	0. 65197	0. 65236	0. 65275	0. 65315
16. 60	. 65551	. 65590	. 65630	. 65669	. 65709
16. 70	. 65945	. 65984	. 66023	. 66063	. 66102
16. 80	. 66338	. 66378	. 66417	. 66457	. 66496
16. 90	. 66732	. 66772	. 66811	. 66850	. 66890
17.00	0. 67126	0. 67165	0. 67205	0. 67244	0. 67283
17. 10	. 67520	. 67559	. 67598	. 67638	. 67677
17. 20	. 67913	. 67953	. 67992	. 68031	. 68071
17. 30	. 68307	. 68346	. 68386	. 68425	. 68464
17. 40	. 68701	. 68740	. 68779	. 68819	. 68858

LENGTH—MILLIMETRES TO INCHES—continued

[From 0.00 to 25.40 millimetres by 0.01 millimetre. 1 millimetre = 0.03937 inch.]

Milli-metres	Hundredths of millimetres				
	0.00	0.01	0.02	0.03	0.04
	Inches	Inches	Inches	Inches	Inches
17.50	0. 68898	0. 68937	0. 68976	0. 69016	0. 69055
17. 60	. 69291	. 69331	. 69370	. 69409	. 69449
17. 70	. 69685	. 69724	. 69764	. 69803	. 69842
17. 80	. 70079	. 70118	. 70157	. 70197	. 70236
17. 90	. 70472	. 70512	. 70551	. 70590	. 70630
18.00	0. 70866	0. 70905	0. 70945	0. 70984	0. 71023
18. 10	. 71260	. 71299	. 71338	. 71378	. 71417
18. 20	. 71653	. 71693	. 71732	. 71772	. 71811
18. 30	. 72047	. 72086	. 72126	. 72165	. 72205
18. 40	. 72441	. 72480	. 72520	. 72559	. 72598
18.50	0. 72834	0. 72874	0. 72913	0. 72953	0. 72992
18. 60	. 73228	. 73268	. 73307	. 73346	. 73386
18. 70	. 73622	. 73661	. 73701	. 73740	. 73779
18. 80	. 74016	. 74055	. 74094	. 74134	. 74173
18. 90	. 74409	. 74449	. 74488	. 74527	. 74567
19.00	0. 74803	0. 74842	0. 74882	0. 74921	0. 74960
19. 10	. 75197	. 75236	. 75275	. 75315	. 75354
19. 20	. 75590	. 75630	. 75669	. 75709	. 75748
19. 30	. 75984	. 76023	. 76063	. 76102	. 76142
19. 40	. 76378	. 76417	. 76457	. 76496	. 76535
19.50	0. 76772	0. 76811	0. 76850	0. 76890	0. 76929
19. 60	. 77165	. 77205	. 77244	. 77283	. 77323
19. 70	. 77559	. 77598	. 77638	. 77677	. 77716
19. 80	. 77953	. 77992	. 78031	. 78071	. 78110
19. 90	. 78346	. 78386	. 78425	. 78464	. 78504
20.00	0. 78740	0. 78779	0. 78819	0. 78858	0. 78897
20. 10	. 79134	. 79173	. 79212	. 79252	. 79291
20. 20	. 79527	. 79567	. 79606	. 79646	. 79685
20. 30	. 79921	. 79960	. 80000	. 80039	. 80079
20. 40	. 80315	. 80354	. 80394	. 80433	. 80472
20.50	0. 80708	0. 80748	0. 80787	0. 80827	0. 80866
20. 60	. 81102	. 81142	. 81181	. 81220	. 81260
20. 70	. 81496	. 81535	. 81575	. 81614	. 81653
20. 80	. 81890	. 81929	. 81968	. 82008	. 82047
20. 90	. 82283	. 82323	. 82362	. 82401	. 82441

LENGTH—MILLIMETRES TO INCHES—continued

[From 0.00 to 25.40 millimetres by 0.01 millimetre. 1 millimetre = 0.03937 inch.]

Milli-metres	Hundredths of millimetres				
	0.05	0.06	0.07	0.08	0.09
	Inches	Inches	Inches	Inches	Inches
17.50	0. 69094	0. 69134	0. 69173	0. 69212	0. 69252
17. 60	. 69488	. 69527	. 69567	. 69606	. 69646
17. 70	. 69882	. 69921	. 69960	. 70000	. 70039
17. 80	. 70275	. 70315	. 70354	. 70394	. 70433
17. 90	. 70669	. 70709	. 70748	. 70787	. 70827
18.00	0. 71063	0. 71102	0. 71142	0. 71181	0. 71220
18. 10	. 71457	. 71496	. 71535	. 71575	. 71614
18. 20	. 71850	. 71890	. 71929	. 71968	. 72008
18. 30	. 72244	. 72283	. 72323	. 72362	. 72401
18. 40	. 72638	. 72677	. 72716	. 72756	. 72795
18.50	0. 73031	0. 73071	0. 73110	0. 73149	0. 73189
18. 60	. 73425	. 73464	. 73504	. 73543	. 73583
18. 70	. 73819	. 73858	. 73897	. 73937	. 73976
18. 80	. 74212	. 74252	. 74291	. 74331	. 74370
18. 90	. 74606	. 74646	. 74685	. 74724	. 74764
19.00	0. 75000	0. 75039	0. 75079	0. 75118	0. 75157
19. 10	. 75394	. 75433	. 75472	. 75512	. 75551
19. 20	. 75787	. 75827	. 75866	. 75905	. 75945
19. 30	. 76181	. 76220	. 76260	. 76299	. 76338
19. 40	. 76575	. 76614	. 76653	. 76693	. 76732
19.50	0. 76968	0. 77008	0. 77047	0. 77086	0. 77126
19. 60	. 77362	. 77401	. 77441	. 77480	. 77520
19. 70	. 77756	. 77795	. 77834	. 77874	. 77913
19. 80	. 78149	. 78189	. 78228	. 78268	. 78307
19. 90	. 78543	. 78583	. 78622	. 78661	. 78701
20.00	0. 78937	0. 78976	0. 79016	0. 79055	0. 79094
20. 10	. 79331	. 79370	. 79409	. 79449	. 79488
20. 20	. 79724	. 79764	. 79803	. 79842	. 79882
20. 30	. 80118	. 80157	. 80197	. 80236	. 80275
20. 40	. 80512	. 80551	. 80590	. 80630	. 80669
20.50	0. 80905	0. 80945	0. 80984	0. 81023	0. 81063
20. 60	. 81299	. 81338	. 81378	. 81417	. 81457
20. 70	. 81693	. 81732	. 81771	. 81811	. 81850
20. 80	. 82086	. 82126	. 82165	. 82205	. 82244
20. 90	. 82480	. 82520	. 82559	. 82598	. 82638

LENGTH—MILLIMETRES TO INCHES—continued

[From 0.00 to 25.40 millimetres by 0.01 millimetre. 1 millimetre = 0.03937 inch.]

Milli-metres	Hundredths of millimetres				
	0.00	0.01	0.02	0.03	0.04
	Inches	Inches	Inches	Inches	Inches
21.00	0. 82677	0. 82716	0. 82756	0. 82795	0. 82834
21. 10	. 83071	. 83110	. 83149	. 83189	. 83228
21. 20	. 83464	. 83504	. 83543	. 83583	. 83622
21. 30	. 83858	. 83897	. 83937	. 83976	. 84016
21. 40	. 84252	. 84291	. 84331	. 84370	. 84409
21.50	0. 84646	0. 84685	0. 84724	0. 84764	0. 84803
21. 60	. 85039	. 85079	. 85118	. 85157	. 85197
21. 70	. 85433	. 85472	. 85512	. 85551	. 85590
21. 80	. 85827	. 85866	. 85905	. 85945	. 85984
21. 90	. 86220	. 86260	. 86299	. 86338	. 86378
22.00	0. 86614	0. 86653	0. 86693	0. 86732	0. 86771
22. 10	. 87008	. 87047	. 87086	. 87126	. 87165
22. 20	. 87401	. 87441	. 87480	. 87520	. 87559
22. 30	. 87795	. 87834	. 87874	. 87913	. 87953
22. 40	. 88189	. 88228	. 88268	. 88307	. 88346
22.50	0. 88582	0. 88622	0. 88661	0. 88701	0. 88740
22. 60	. 88976	. 89016	. 89055	. 89094	. 89134
22. 70	. 89370	. 89409	. 89449	. 89488	. 89527
22. 80	. 89764	. 89803	. 89842	. 89882	. 89921
22. 90	. 90157	. 90197	. 90236	. 90275	. 90315
23.00	0. 90551	0. 90590	0. 90630	0. 90669	0. 90708
23. 10	. 90945	. 90984	. 91023	. 91063	. 91102
23. 20	. 91338	. 91378	. 91417	. 91457	. 91496
23. 30	. 91732	. 91771	. 91811	. 91850	. 91890
23. 40	. 92126	. 92165	. 92205	. 92244	. 92283
23.50	0. 92520	0. 92559	0. 92598	0. 92638	0. 92677
23. 60	. 92913	. 92953	. 92992	. 93031	. 93071
23. 70	. 93307	. 93346	. 93386	. 93425	. 93464
23. 80	. 93701	. 93740	. 93779	. 93819	. 93858
23. 90	. 94094	. 94134	. 94173	. 94212	. 94252
24.00	0. 94488	0. 94527	0. 94567	0. 94606	0. 94645
24. 10	. 94882	. 94921	. 94960	. 95000	. 95039
24. 20	. 95275	. 95315	. 95354	. 95394	. 95433
24. 30	. 95669	. 95708	. 95748	. 95787	. 95827
24. 40	. 96063	. 96102	. 96142	. 96181	. 96220

LENGTH—MILLIMETRES TO INCHES—continued

[From 0.00 to 25.40 millimetres by 0.01 millimetre. 1 millimetre = 0.03937 inch.]

Milli-metres	Hundredths of millimetres				
	0.05	0.06	0.07	0.08	0.09
	Inches	Inches	Inches	Inches	Inches
21.00	0. 82874	0. 82913	0. 82953	0. 82992	0. 83031
21. 10	. 83268	. 83307	. 83346	. 83386	. 83425
21. 20	. 83661	. 83701	. 83740	. 83779	. 83819
21. 30	. 84055	. 84094	. 84134	. 84173	. 84212
21. 40	. 84449	. 84488	. 84527	. 84567	. 84606
21.50	0. 84842	0. 84882	0. 84921	0. 84960	0. 85000
21. 60	. 85236	. 85275	. 85315	. 85354	. 85394
21. 70	. 85630	. 85669	. 85708	. 85748	. 85787
21. 80	. 86023	. 86063	. 86102	. 86142	. 86181
21. 90	. 86417	. 86457	. 86496	. 86535	. 86575
22.00	0. 86811	0. 86850	0. 86890	0. 86929	0. 86968
22. 10	. 87205	. 87244	. 87283	. 87323	. 87362
22. 20	. 87598	. 87638	. 87677	. 87716	. 87756
22. 30	. 87992	. 88031	. 88071	. 88110	. 88149
22. 40	. 88386	. 88425	. 88464	. 88504	. 88543
22.50	0. 88779	0. 88819	0. 88858	0. 88897	0. 88937
22. 60	. 89173	. 89212	. 89252	. 89291	. 89331
22. 70	. 89567	. 89606	. 89645	. 89685	. 89724
22. 80	. 89960	. 90000	. 90039	. 90079	. 90118
22. 90	. 90354	. 90394	. 90433	. 90472	. 90512
23.00	0. 90748	0. 90787	0. 90827	0. 90866	0. 90905
23. 10	. 91142	. 91181	. 91220	. 91260	. 91299
23. 20	. 91535	. 91575	. 91614	. 91653	. 91693
23. 30	. 91929	. 91968	. 92008	. 92047	. 92086
23. 40	. 92323	. 92362	. 92401	. 92441	. 92480
23.50	0. 92716	0. 92756	0. 92795	0. 92834	0. 92874
23. 60	. 93110	. 93149	. 93189	. 93228	. 93268
23. 70	. 93504	. 93543	. 93582	. 93622	. 93661
23. 80	. 93897	. 93937	. 93976	. 94016	. 94055
23. 90	. 94291	. 94331	. 94370	. 94409	. 94449
24.00	0. 94685	0. 94724	0. 94764	0. 94803	0. 94842
24. 10	. 95079	. 95118	. 95157	. 95197	. 95236
24. 20	. 95472	. 95512	. 95551	. 95590	. 95630
24. 30	. 95866	. 95905	. 95945	. 95984	. 96023
24. 40	. 96260	. 96299	. 96338	. 96378	. 96417

LENGTH—MILLIMETRES TO INCHES—continued

[From 0.00 to 25.40 millimetres by 0.01 millimetre. 1 millimetre = 0.03937 inch.]

Milli-metres	Hundredths of millimetres				
	0.00	0.01	0.02	0.03	0.04
	Inches	Inches	Inches	Inches	Inches
24.50	0.96456	0.96496	0.96535	0.96575	0.96614
24.60	.96850	.96890	.96929	.96968	.97008
24.70	.97244	.97283	.97323	.97362	.97401
24.80	.97638	.97677	.97716	.97756	.97795
24.90	.98031	.98071	.98110	.98149	.98189
25.00	0.98425	0.98464	0.98504	0.98543	0.98582
25.10	.98819	.98858	.98897	.98937	.98976
25.20	.99212	.99252	.99291	.99331	.99370
25.30	.99606	.99645	.99685	.99724	.99764
25.40	1.00000				

```
 1 inch   = 0.02540 metre
 2 inches =  .05080 metre
 3 inches =  .07620 metre
 4 inches = 0.10160 metre
 5 inches =  .12700 metre
 6 inches =  .15240 metre
 7 inches = 0.17780 metre
 8 inches =  .20320 metre
 9 inches =  .22860 metre
10 inches = 0.25400 metre
11 inches =  .27940 metre
12 inches =  .30480 metre
```

LENGTH—MILLIMETRES TO INCHES—continued

[From 0.00 to 25.40 millimetres by 0.01 millimetre. 1 millimetre = 0.03937 inch.]

Milli-metres	Hundredths of millimetres				
	0.05	0.06	0.07	0.08	0.09
	Inches	Inches	Inches	Inches	Inches
24.50	0.96653	0.96693	0.96732	0.96771	0.96811
24.60	.97047	.97086	.97126	.97165	.97205
24.70	.97441	.97480	.97519	.97559	.97598
24.80	.97834	.97874	.97913	.97953	.97992
24.90	.98228	.98268	.98307	.98346	.98386
25.00	0.98622	0.98661	0.98701	0.98740	0.98779
25.10	.99016	.99055	.99094	.99134	.99173
25.20	.99409	.99449	.99488	.99527	.99567
25.30	.99803	.99842	.99882	.99921	.99960
25.40	....	....	....	....	....

The above tables converting millimetres to inches may be used to convert centimetres to inches by moving the decimal point 1 place to the right, and decimeters to inches by moving the decimal point 2 places to the right.

EXAMPLE: 1 millimetre = 0.03937 inches
1 centimetre = 0.3937   ″
1 decimetre = 3.937   ″
1 metre   = 39.37   ″

LENGTH—METRES TO FEET

[Reduction factor: 1 metre = 3.280833333 feet]

Metres	Feet	Metres	Feet	Metres	Feet	Metres	Feet	Metres	Feet
0		50	164.04167	100	328.08333	150	492.12500	200	656.16667
1	3.28083	1	167.32250	1	331.36417	1	495.40583	1	659.44750
2	6.56167	2	170.60333	2	334.64500	2	498.68667	2	662.72833
3	9.84250	3	173.88417	3	337.92583	3	501.96750	3	666.00917
4	13.12333	4	177.16500	4	341.20667	4	505.24833	4	669.29000
5	16.40417	5	180.44583	5	344.48750	5	508.52917	5	672.57083
6	19.68500	6	183.72667	6	347.76833	6	511.81000	6	675.85167
7	22.96583	7	187.00750	7	351.04917	7	515.09083	7	679.13250
8	26.24667	8	190.28833	8	354.33000	8	518.37167	8	682.41333
9	29.52750	9	193.56917	9	357.61083	9	521.65250	9	685.69417
10	32.80833	60	196.85000	110	360.89167	160	524.93333	210	688.97500
1	36.08917	1	200.13083	1	364.17250	1	528.21417	1	692.25583
2	39.37000	2	203.41167	2	367.45333	2	531.49500	2	695.53667
3	42.65083	3	206.69250	3	370.73417	3	534.77583	3	698.81750
4	45.93167	4	209.97333	4	374.01500	4	538.05667	4	702.09833
5	49.21250	5	213.25417	5	377.29583	5	541.33750	5	705.37917
6	52.49333	6	216.53500	6	380.57667	6	544.61833	6	708.66000
7	55.77417	7	219.81583	7	383.85750	7	547.89917	7	711.94083
8	59.05500	8	223.09667	8	387.13833	8	551.18000	8	715.22167
9	62.33583	9	226.37750	9	390.41917	9	554.46083	9	718.50250
20	65.61667	70	229.65833	120	393.70000	170	557.74167	220	721.78333
1	68.89750	1	232.93917	1	396.98083	1	561.02250	1	725.06417
2	72.17833	2	236.22000	2	400.26167	2	564.30333	2	728.34500
3	75.45917	3	239.50083	3	403.54250	3	567.58417	3	731.62583
4	78.74000	4	242.78167	4	406.82333	4	570.86500	4	734.90667
5	82.02083	5	246.06250	5	410.10417	5	574.14583	5	738.18750
6	85.30167	6	249.34333	6	413.38500	6	577.42667	6	741.46833
7	88.58250	7	252.62417	7	416.66583	7	580.70750	7	744.74917
8	91.86333	8	255.90500	8	419.94667	8	583.98833	8	748.03000
9	95.14417	9	259.18583	9	423.22750	9	587.26917	9	751.31083
30	98.42500	80	262.46667	130	426.50833	180	590.55000	230	754.59167
1	101.70583	1	265.74750	1	429.78917	1	593.83083	1	757.87250
2	104.98667	2	269.02833	2	433.07000	2	597.11167	2	761.15333
3	108.26750	3	272.30917	3	436.35083	3	600.39250	3	764.43417
4	111.54833	4	275.59000	4	439.63167	4	603.67333	4	767.71500
5	114.82917	5	278.87083	5	442.91250	5	606.95417	5	770.99583
6	118.11000	6	282.15167	6	446.19333	6	610.23500	6	774.27667
7	121.39083	7	285.43250	7	449.47417	7	613.51583	7	777.55750
8	124.67167	8	288.71333	8	452.75500	8	616.79667	8	780.83833
9	127.95250	9	291.99417	9	456.03583	9	620.07750	9	784.11917
40	131.23333	90	295.27500	140	459.31667	190	623.35833	240	787.40000
1	134.51417	1	298.55583	1	462.59750	1	626.63917	1	790.68083
2	137.79500	2	301.83667	2	465.87833	2	629.92000	2	793.96167
3	141.07583	3	305.11750	3	469.15917	3	633.20083	3	797.24250
4	144.35667	4	308.39833	4	472.44000	4	636.48167	4	800.52333
5	147.63750	5	311.67917	5	475.72083	5	639.76250	5	803.80417
6	150.91833	6	314.96000	6	479.00167	6	643.04333	6	807.08500
7	154.19917	7	318.24083	7	482.28250	7	646.32417	7	810.36583
8	157.48000	8	321.52167	8	485.56333	8	649.60500	8	813.64667
9	160.76083	9	324.80250	9	488.84417	9	652.88583	9	816.92750

LENGTH—METRES TO FEET

[Reduction factor: 1 metre = 3.280833333 feet]

Metres	Feet	Metres	Feet	Metres	Feet	Metres	Feet	Metres	Feet
250	820.20833	300	984.25000	350	1,148.29167	400	1,312.33333	450	1,476.37500
1	823.48917	1	987.53083	1	1,151.57250	1	1,315.61417	1	1,479.65583
2	826.77000	2	990.81167	2	1,154.85333	2	1,318.89500	2	1,482.93667
3	830.05083	3	994.09250	3	1,158.13417	3	1,322.17583	3	1,486.21750
4	833.33167	4	997.37333	4	1,161.41500	4	1,325.45667	4	1,489.49833
5	836.61250	5	1,000.65417	5	1,164.69583	5	1,328.73750	5	1,492.77917
6	839.89333	6	1,003.93500	6	1,167.97667	6	1,332.01833	6	1,496.06000
7	843.17417	7	1,007.21583	7	1,171.25750	7	1,335.29917	7	1,499.34083
8	846.45500	8	1,010.49667	8	1,174.53833	8	1,338.58000	8	1,502.62167
9	849.73583	9	1,013.77750	9	1,177.81917	9	1,341.86083	9	1,505.90250
260	853.01667	310	1,017.05833	360	1,181.10000	410	1,345.14167	460	1,509.18333
1	856.29750	1	1,020.33917	1	1,184.38083	1	1,348.42250	1	1,512.46417
2	859.57833	2	1,023.62000	2	1,187.66167	2	1,351.70333	2	1,515.74500
3	862.85917	3	1,026.90083	3	1,190.94250	3	1,354.98417	3	1,519.02583
4	866.14000	4	1,030.18167	4	1,194.22333	4	1,358.26500	4	1,522.30667
5	869.42083	5	1,033.46250	5	1,197.50417	5	1,361.54583	5	1,525.58750
6	872.70167	6	1,036.74333	6	1,200.78500	6	1,364.82667	6	1,528.86833
7	875.98250	7	1,040.02417	7	1,204.06583	7	1,368.10750	7	1,532.14917
8	879.26333	8	1,043.30500	8	1,207.34667	8	1,371.38833	8	1,535.43000
9	882.54417	9	1,046.58583	9	1,210.62750	9	1,374.66917	9	1,538.71083
270	885.82500	320	1,049.86667	370	1,213.90833	420	1,377.95000	470	1,541.99167
1	889.10583	1	1,053.14750	1	1,217.18917	1	1,381.23083	1	1,545.27250
2	892.38667	2	1,056.42833	2	1,220.47000	2	1,384.51167	2	1,548.55333
3	895.66750	3	1,059.70917	3	1,223.75083	3	1,387.79250	3	1,551.83417
4	898.94833	4	1,062.99000	4	1,227.03167	4	1,391.07333	4	1,555.11500
5	902.22917	5	1,066.27083	5	1,230.31250	5	1,394.35417	5	1,558.39583
6	905.51000	6	1,069.55167	6	1,233.59353	6	1,397.63500	6	1,561.67667
7	908.79083	7	1,072.83250	7	1,236.87417	7	1,400.91583	7	1,564.95750
8	912.07167	8	1,076.11333	8	1,240.15500	8	1,404.19667	8	1,568.23833
9	915.35250	9	1,079.39417	9	1,243.43583	9	1,407.47750	9	1,571.51917
280	918.63333	330	1,082.67500	380	1,246.71667	430	1,410.75833	480	1,574.80000
1	921.91417	1	1,085.95583	1	1,249.99750	1	1,414.03917	1	1,578.08083
2	925.19500	2	1,089.23667	2	1,253.27833	2	1,417.32000	2	1,581.36167
3	928.47583	3	1,092.51750	3	1,256.55917	3	1,420.60083	3	1,584.64250
4	931.75667	4	1,095.79833	4	1,259.84000	4	1,423.88167	4	1,587.92333
5	935.03750	5	1,099.07917	5	1,263.12083	5	1,427.16250	5	1,591.20417
6	938.31833	6	1,102.36000	6	1,266.40167	6	1,430.44333	6	1,594.48500
7	941.59917	7	1,105.64083	7	1,269.68250	7	1,433.72417	7	1,597.76583
8	944.88000	8	1,108.92167	8	1,272.96333	8	1,437.00500	8	1,601.04667
9	948.16083	9	1,112.20250	9	1,276.24417	9	1,440.28583	9	1,604.32750
290	951.44167	340	1,115.48333	390	1,279.52500	440	1,443.56667	490	1,607.60833
1	954.72250	1	1,118.76417	1	1,282.80583	1	1,446.84750	1	1,610.88917
2	958.00333	2	1,122.04500	2	1,286.08667	2	1,450.12833	2	1,614.17000
3	961.28417	3	1,125.32583	3	1,289.36750	3	1,453.40917	3	1,617.45083
4	964.56500	4	1,128.60667	4	1,292.64833	4	1,456.69000	4	1,620.73167
5	967.84583	5	1,131.88750	5	1,295.92917	5	1,459.97083	5	1,624.01250
6	971.12667	6	1,135.16833	6	1,299.21000	6	1,463.25167	6	1,627.29333
7	974.40750	7	1,138.44917	7	1,302.49083	7	1,466.53250	7	1,630.57417
8	977.68833	8	1,141.73000	8	1,305.77167	8	1,469.81333	8	1,633.85500
9	980.96917	9	1,145.01083	9	1,309.05250	9	1,473.09417	9	1,637.13583

LENGTH—METRES TO FEET

[Reduction factor: 1 metre = 3.280833333 feet]

Metres	Feet	Metres	Feet	Metres	Feet	Metres	Feet	Metres	Feet
500	1,640.41667	550	1,804.45833	600	1,968.50000	650	2,132.54167	700	2,296.58333
1	1,643.69750	1	1,807.73917	1	1,971.78083	1	2,135.82250	1	2,299.86417
2	1,646.97833	2	1,811.02000	2	1,975.06167	2	2,139.10333	2	2,303.14500
3	1,650.25917	3	1,814.30083	3	1,978.34250	3	2,142.38417	3	2,306.42583
4	1,653.54000	4	1,817.58167	4	1,981.62333	4	2,145.66500	4	2,309.70667
5	1,656.82083	5	1,820.86250	5	1,984.90417	5	2,148.94583	5	2,312.98750
6	1,660.10167	6	1,824.14333	6	1,988.18500	6	2,152.22667	6	2,316.26833
7	1,663.38250	7	1,827.42417	7	1,991.46583	7	2,155.50750	7	2,319.54917
8	1,666.66333	8	1,830.70500	8	1,994.74667	8	2,158.78833	8	2,322.83000
9	1,669.94417	9	1,833.98583	9	1,998.02750	9	2,162.06917	9	2,326.11083
510	1,673.22500	560	1,837.26667	610	2,001.30833	660	2,165.35000	710	2,329.39167
1	1,676.50583	1	1,840.54750	1	2,004.58917	1	2,168.63083	1	2,332.67250
2	1,679.78667	2	1,843.82833	2	2,007.87000	2	2,171.91167	2	2,335.95333
3	1,683.06750	3	1,847.10917	3	2,011.15083	3	2,175.19250	3	2,339.23417
4	1,686.34833	4	1,850.39000	4	2,014.43167	4	2,178.47333	4	2,342.51500
5	1,689.62917	5	1,853.67083	5	2,017.71250	5	2,181.75417	5	2,345.79583
6	1,692.91000	6	1,856.95167	6	2,020.99333	6	2,185.03500	6	2,349.07667
7	1,696.19083	7	1,860.23250	7	2,024.27417	7	2,188.31583	7	2,352.35750
8	1,699.47167	8	1,863.51333	8	2,027.55500	8	2,191.59667	8	2,355.63833
9	1,702.75250	9	1,866.79417	9	2,030.83583	9	2,194.87750	9	2,358.91917
520	1,706.03333	570	1,870.07500	620	2,034.11667	670	2,198.15833	720	2,362.20000
1	1,709.31417	1	1,873.35583	1	2,037.39750	1	2,201.43917	1	2,365.48083
2	1,712.59500	2	1,876.63667	2	2,040.67833	2	2,204.72000	2	2,368.76167
3	1,715.87583	3	1,879.91750	3	2,043.95917	3	2,208.00083	3	2,372.04250
4	1,719.15667	4	1,883.19833	4	2,047.24000	4	2,211.28167	4	2,375.32333
5	1,722.43750	5	1,886.47917	5	2,050.52083	5	2,214.56250	5	2,378.60417
6	1,725.71833	6	1,889.76000	6	2,053.80167	6	2,217.84333	6	2,381.88500
7	1,728.99917	7	1,893.04083	7	2,057.08250	7	2,221.12417	7	2,385.16583
8	1,732.28000	8	1,896.32167	8	2,060.36333	8	2,224.40500	8	2,388.44667
9	1,735.56083	9	1,899.60250	9	2,063.64417	9	2,227.68583	9	2,391.72750
530	1,738.84167	580	1,902.88333	630	2,066.92500	680	2,230.96667	730	2,395.00833
1	1,742.12250	1	1,906.16417	1	2,070.20583	1	2,234.24750	1	2,398.28917
2	1,745.40333	2	1,909.44500	2	2,073.48667	2	2,237.52833	2	2,401.57000
3	1,748.68417	3	1,912.72583	3	2,076.76750	3	2,240.80917	3	2,404.85083
4	1,751.96500	4	1,916.00667	4	2,080.04833	4	2,244.09000	4	2,408.13167
5	1,755.24583	5	1,919.28750	5	2,083.32917	5	2,247.37083	5	2,411.41250
6	1,758.52667	6	1,922.56833	6	2,086.61000	6	2,250.65167	6	2,414.69333
7	1,761.80750	7	1,925.84917	7	2,089.89083	7	2,253.93250	7	2,417.97417
8	1,765.08833	8	1,929.13000	8	2,093.17167	8	2,257.21333	8	2,421.25500
9	1,768.36917	9	1,932.41083	9	2,096.45250	9	2,260.49417	9	2,424.53583
540	1,771.65000	590	1,935.69167	640	2,099.73333	690	2,263.77500	740	2,427.81667
1	1,774.93083	1	1,938.97250	1	2,103.01417	1	2,267.05583	1	2,431.09750
2	1,778.21167	2	1,942.25333	2	2,106.29500	2	2,270.33667	2	2,434.37833
3	1,781.49250	3	1,945.53417	3	2,109.57583	3	2,273.61750	3	2,437.65917
4	1,784.77333	4	1,948.81500	4	2,112.85667	4	2,276.89833	4	2,440.94000
5	1,788.05417	5	1,952.09583	5	2,116.13750	5	2,280.17917	5	2,444.22083
6	1,791.33500	6	1,955.37667	6	2,119.41833	6	2,283.46000	6	2,447.50167
7	1,794.61583	7	1,958.65750	7	2,122.69917	7	2,286.74083	7	2,450.78250
8	1,797.89667	8	1,961.93833	8	2,125.98000	8	2,290.02167	8	2,454.06333
9	1,801.17750	9	1,965.21917	9	2,129.26083	9	2,293.30250	9	2,457.34417

LENGTH—METRES TO FEET

[Reduction factor: 1 metre = 3.280833333 feet]

Metres	Feet	Metres	Feet	Metres	Feet	Metres	Feet	Metres	Feet
750	2,460.62500	800	2,624.66667	850	2,788.70833	900	2,952.75000	950	3,116.79167
1	2,463.90583	1	2,627.94750	1	2,791.98917	1	2,956.03083	1	3,120.07250
2	2,467.18667	2	2,631.22833	2	2,795.27000	2	2,959.31167	2	3,123.35333
3	2,470.46750	3	2,634.50917	3	2,798.55083	3	2,962.59250	3	3,126.63417
4	2,473.74833	4	2,637.79000	4	2,801.83167	4	2,965.87333	4	3,129.91500
5	2,477.02917	5	2,641.07083	5	2,805.11250	5	2,969.15417	5	3,133.19583
6	2,480.31000	6	2,644.35167	6	2,808.39333	6	2,972.43500	6	3,136.47667
7	2,483.59083	7	2,647.63250	7	2,811.67417	7	2,975.71583	7	3,139.75750
8	2,486.87167	8	2,650.91333	8	2,814.95500	8	2,978.99667	8	3,143.03833
9	2,490.15250	9	2,654.19417	9	2,818.23583	9	2,982.27750	9	3,146.31917
760	2,493.43333	810	2,657.47500	860	2,821.51667	910	2,985.55833	960	3,149.60000
1	2,496.71417	1	2,660.75583	1	2,824.79750	1	2,988.83917	1	3,152.88083
2	2,499.99500	2	2,664.03667	2	2,828.07833	2	2,992.12000	2	3,156.16167
3	2,503.27583	3	2,667.31750	3	2,831.35917	3	2,995.40083	3	3,159.44250
4	2,506.55667	4	2,670.59833	4	2,834.64000	4	2,998.68167	4	3,162.72333
5	2,509.83750	5	2,673.87917	5	2,837.92083	5	3,001.96250	5	3,166.00417
6	2,513.11833	6	2,677.16000	6	2,841.20167	6	3,005.24333	6	3,169.28500
7	2,516.39917	7	2,680.44083	7	2,844.48250	7	3,008.52417	7	3,172.56583
8	2,519.68000	8	2,683.72167	8	2,847.76333	8	3,011.80500	8	3,175.84667
9	2,522.96083	9	2,687.00250	9	2,851.04417	9	3,015.08583	9	3,179.12750
770	2,526.24167	820	2,690.28333	870	2,854.32500	920	3,018.36667	970	3,182.40833
1	2,529.52250	1	2,693.56417	1	2,857.60583	1	3,021.64750	1	3,185.68917
2	2,532.80333	2	2,696.84500	2	2,860.88667	2	3,024.92833	2	3,188.97000
3	2,536.08417	3	2,700.12583	3	2,864.16750	3	3,028.20917	3	3,192.25083
4	2,539.36500	4	2,703.40667	4	2,867.44833	4	3,031.49000	4	3,195.53167
5	2,542.64583	5	2,706.68750	5	2,870.72917	5	3,034.77083	5	3,198.81250
6	2,545.92667	6	2,709.96833	6	2,874.01000	6	3,038.05167	6	3,202.09333
7	2,549.20750	7	2,713.24917	7	2,877.29083	7	3,041.33250	7	3,205.37417
8	2,552.48833	8	2,716.53000	8	2,880.57167	8	3,044.61333	8	3,208.65500
9	2,555.76917	9	2,719.81083	9	2,883.85250	9	3,047.89417	9	3,211.93583
780	2,559.05000	830	2,723.09167	880	2,887.13333	930	3,051.17500	980	3,215.21667
1	2,562.33083	1	2,726.37250	1	2,890.41417	1	3,054.45583	1	3,218.49750
2	2,565.61167	2	2,729.65333	2	2,893.69500	2	3,057.73667	2	3,221.77833
3	2,568.89250	3	2,732.93417	3	2,896.97583	3	3,061.01750	3	3,225.05917
4	2,572.17333	4	2,736.21500	4	2,900.25667	4	3,064.29833	4	3,228.34000
5	2,575.45417	5	2,739.49583	5	2,903.53750	5	3,067.57917	5	3,231.62083
6	2,578.73500	6	2,742.77667	6	2,906.81833	6	3,070.86000	6	3,234.90167
7	2,582.01583	7	2,746.05750	7	2,910.09917	7	3,074.14083	7	3,238.18250
8	2,585.29667	8	2,749.33833	8	2,913.38000	8	3,077.42167	8	3,241.46333
9	2,588.57750	9	2,752.61917	9	2,916.66083	9	3,080.70250	9	3,244.74417
790	2,591.85833	840	2,755.90000	890	2,919.94167	940	3,083.98333	990	3,248.02500
1	2,595.13917	1	2,759.18083	1	2,923.22250	1	3,087.26417	1	3,251.30583
2	2,598.42000	2	2,762.46167	2	2,926.50333	2	3,090.54500	2	3,254.58667
3	2,601.70083	3	2,765.74250	3	2,929.78417	3	3,093.82583	3	3,257.86750
4	2,604.98167	4	2,769.02333	4	2,933.06500	4	3,097.10667	4	3,261.14833
5	2,608.26250	5	2,772.30417	5	2,936.34583	5	3,100.38750	5	3,264.42917
6	2,611.54333	6	2,775.58500	6	2,939.62667	6	3,103.66833	6	3,267.71000
7	2,614.82417	7	2,778.86583	7	2,942.90750	7	3,106.94917	7	3,270.99083
8	2,618.10500	8	2,782.14667	8	2,946.18833	8	3,110.23000	8	3,274.27167
9	2,621.38583	9	2,785.42750	9	2,949.46917	9	3,113.51083	9	3,277.55250

LENGTH—FEET TO METRES

[Reduction factor: 1 foot = 0.3048006096 metre]

Feet	Metres	Feet	Metres	Feet	Metres	Feet	Metres	Feet	Metres
0		50	15.24003	100	30.48006	150	45.72009	200	60.96012
1	0.30480	1	15.54483	1	30.78486	1	46.02489	1	61.26492
2	.60960	2	15.84963	2	31.08966	2	46.32969	2	61.56972
3	.91440	3	16.15443	3	31.39446	3	46.63449	3	61.87452
4	1.21920	4	16.45923	4	31.69926	4	46.93929	4	62.17932
5	1.52400	5	16.76403	5	32.00406	5	47.24409	5	62.48412
6	1.82880	6	17.06883	6	32.30886	6	47.54890	6	62.78893
7	2.13360	7	17.37363	7	32.61367	7	47.85370	7	63.09373
8	2.43840	8	17.67844	8	32.91847	8	48.15850	8	63.39853
9	2.74321	9	17.98324	9	33.22327	9	48.46330	9	63.70333
10	3.04801	60	18.28804	110	33.52807	160	48.76810	210	64.00813
1	3.35281	1	18.59284	1	33.83287	1	49.07290	1	64.31293
2	3.65761	2	18.89764	2	34.13767	2	49.37770	2	64.61773
3	3.96241	3	19.20244	3	34.44247	3	49.68250	3	64.92253
4	4.26721	4	19.50724	4	34.74727	4	49.98730	4	65.22733
5	4.57201	5	19.81204	5	35.05207	5	50.29210	5	65.53213
6	4.87681	6	20.11684	6	35.35687	6	50.59690	6	65.83693
7	5.18161	7	20.42164	7	35.66167	7	50.90170	7	66.14173
8	5.48641	8	20.72644	8	35.96647	8	51.20650	8	66.44653
9	5.79121	9	21.03124	9	36.27127	9	51.51130	9	66.75133
20	6.09601	70	21.33604	120	36.57607	170	51.81610	220	67.05613
1	6.40081	1	21.64084	1	36.88087	1	52.12090	1	67.36093
2	6.70561	2	21.94564	2	37.18567	2	52.42570	2	67.66574
3	7.01041	3	22.25044	3	37.49047	3	52.73051	3	67.97054
4	7.31521	4	22.55525	4	37.79528	4	53.03531	4	68.27534
5	7.62002	5	22.86005	5	38.10008	5	53.34011	5	68.58014
6	7.92482	6	23.16485	6	38.40488	6	53.64491	6	68.88494
7	8.22962	7	23.46965	7	38.70968	7	53.94971	7	69.18974
8	8.53442	8	23.77445	8	39.01448	8	54.25451	8	69.49454
9	8.83922	9	24.07925	9	39.31928	9	54.55931	9	69.79934
30	9.14402	80	24.38405	130	39.62408	180	54.86411	230	70.10414
1	9.44882	1	24.68885	1	39.92888	1	55.16891	1	70.40894
2	9.75362	2	24.99365	2	40.23368	2	55.47371	2	70.71374
3	10.05842	3	25.29845	3	40.53848	3	55.77851	3	71.01854
4	10.36322	4	25.60325	4	40.84328	4	56.08331	4	71.32334
5	10.66802	5	25.90805	5	41.14808	5	56.38811	5	71.62814
6	10.97383	6	26.21285	6	41.45288	6	56.69291	6	71.93294
7	11.27762	7	26.51765	7	41.75768	7	56.99771	7	72.23774
8	11.58242	8	26.82245	8	42.06248	8	57.30251	8	72.54255
9	11.88722	9	27.12725	9	42.36728	9	57.60732	9	72.84735
40	12.19202	90	27.43205	140	42.67209	190	57.91212	240	73.15215
1	12.49682	1	27.73686	1	42.97689	1	58.21692	1	73.45695
2	12.80163	2	28.04166	2	43.28169	2	58.52172	2	73.76175
3	13.10643	3	28.34646	3	43.58649	3	58.82652	3	74.06655
4	13.41123	4	28.65126	4	43.89129	4	59.13132	4	74.37135
5	13.71603	5	28.95606	5	44.19609	5	59.43612	5	74.67615
6	14.02083	6	29.26086	6	44.50089	6	59.74092	6	74.98095
7	14.32563	7	29.56566	7	44.80569	7	60.04572	7	75.28575
8	14.63043	8	29.87046	8	45.11049	8	60.35052	8	75.59055
9	14.93523	9	30.17526	9	45.41529	9	60.65532	9	75.89535

LENGTH—FEET TO METRES

[Reduction factor: 1 foot = 0.3048006096 metre]

Feet	Metres	Feet	Metres	Feet	Metres	Feet	Metres	Feet	Metres
250	76.20015	300	91.44018	350	106.68021	400	121.92024	450	137.16027
1	76.50495	1	91.74498	1	106.98501	1	122.22504	1	137.46507
2	76.80975	2	92.04978	2	107.28981	2	122.52985	2	137.76988
3	77.11455	3	92.35458	3	107.59462	3	122.83465	3	138.07468
4	77.41935	4	92.65939	4	107.89942	4	123.13945	4	138.37948
5	77.72416	5	92.96419	5	108.20422	5	123.44425	5	138.68428
6	78.02896	6	93.26899	6	108.50902	6	123.74905	6	138.98908
7	78.33376	7	93.57379	7	108.81382	7	124.05385	7	139.29388
8	78.63856	8	93.87859	8	109.11862	8	124.35865	8	139.59868
9	78.94336	9	94.18339	9	109.42342	9	124.66345	9	139.90348
260	79.24816	310	94.48819	360	109.72822	410	124.96825	460	140.20828
1	79.55296	1	94.79299	1	110.03302	1	125.27305	1	140.51308
2	79.85776	2	95.09779	2	110.33782	2	125.57785	2	140.81788
3	80.16256	3	95.40259	3	110.64262	3	125.88265	3	141.12268
4	80.46736	4	95.70739	4	110.94742	4	126.18745	4	141.42748
5	80.77216	5	96.01219	5	111.25222	5	126.49225	5	141.73228
6	81.07696	6	96.31699	6	111.55702	6	126.79705	6	142.03708
7	81.38176	7	96.62179	7	111.86182	7	127.10185	7	142.34188
8	81.68656	8	96.92659	8	112.16662	8	127.40665	8	142.64669
9	81.99136	9	97.23139	9	112.47142	9	127.71146	9	142.95149
270	82.29616	320	97.53620	370	112.77623	420	128.01626	470	143.25629
1	82.60097	1	97.84100	1	113.08103	1	128.32106	1	143.56109
2	82.90577	2	98.14580	2	113.38583	2	128.62586	2	143.86589
3	83.21057	3	98.45060	3	113.69063	3	128.93066	3	144.17069
4	83.51537	4	98.75540	4	113.99543	4	129.23546	4	144.47549
5	83.82017	5	99.06020	5	114.30023	5	129.54026	5	144.78029
6	84.12497	6	99.36500	6	114.60503	6	129.84506	6	145.08509
7	84.42977	7	99.66980	7	114.90983	7	130.14986	7	145.38989
8	84.73457	8	99.97460	8	115.21463	8	130.45466	8	145.69469
9	85.03937	9	100.27940	9	115.51943	9	130.75946	9	145.99949
280	85.34417	330	100.58420	380	115.82423	430	131.06426	480	146.30429
1	85.64897	1	100.88900	1	116.12903	1	131.36906	1	146.60909
2	85.95377	2	101.19380	2	116.43383	2	131.67386	2	146.91389
3	86.25857	3	101.49860	3	116.73863	3	131.97866	3	147.21869
4	86.56337	4	101.80340	4	117.04343	4	132.28346	4	147.52350
5	86.86817	5	102.10820	5	117.34823	5	132.58827	5	147.82830
6	87.17297	6	102.41300	6	117.65304	6	132.89307	6	148.13310
7	87.47777	7	102.71781	7	117.95784	7	133.19787	7	148.43790
8	87.78258	8	103.02261	8	118.26264	8	133.50267	8	148.74270
9	88.08738	9	103.32741	9	118.56744	9	133.80747	9	149.04750
290	88.39218	340	103.63221	390	118.87224	440	134.11227	490	149.35230
1	88.69698	1	103.93701	1	119.17704	1	134.41707	1	149.65710
2	89.00178	2	104.24181	2	119.48184	2	134.72187	2	149.96190
3	89.30658	3	104.54661	3	119.78664	3	135.02667	3	150.26670
4	89.61138	4	104.85141	4	120.09144	4	135.33147	4	150.57150
5	89.91618	5	105.15621	5	120.39624	5	135.63627	5	150.87630
6	90.22098	6	105.46101	6	120.70104	6	135.94107	6	151.18110
7	90.52578	7	105.76581	7	121.00584	7	136.24587	7	151.48590
8	90.83058	8	106.07061	8	121.31064	8	136.55067	8	151.79070
9	91.13538	9	106.37541	9	121.61544	9	136.85547	9	152.09550

LENGTH—FEET TO METRES

[Reduction factor: 1 foot = 0.3048006096 metre]

Feet	Metres	Feet	Metres	Feet	Metres	Feet	Metres	Feet	Metres
500	152.40030	550	167.64034	600	182.88037	650	198.12040	700	213.36043
1	152.70511	1	167.94514	1	183.18517	1	198.42520	1	213.66523
2	153.00991	2	168.24994	2	183.48997	2	198.73000	2	213.97003
3	153.31471	3	168.55474	3	183.79477	3	199.03480	3	214.27483
4	153.61951	4	168.85954	4	184.09957	4	199.33960	4	214.57963
5	153.92431	5	169.16434	5	184.40437	5	199.64440	5	214.88443
6	154.22911	6	169.46914	6	184.70917	6	199.94920	6	215.18923
7	154.53391	7	169.77394	7	185.01397	7	200.25400	7	215.49403
8	154.83871	8	170.07874	8	185.31877	8	200.55880	8	215.79883
9	155.14351	9	170.38354	9	185.62357	9	200.86360	9	216.10363
510	155.44831	560	170.68834	610	185.92837	660	201.16840	710	216.40843
1	155.75311	1	170.99314	1	186.23317	1	201.47320	1	216.71323
2	156.05791	2	171.29794	2	186.53797	2	201.77800	2	217.01803
3	156.36271	3	171.60274	3	186.84277	3	202.08280	3	217.32283
4	156.66751	4	171.90754	4	187.14757	4	202.38760	4	217.62764
5	156.97231	5	172.21234	5	187.45237	5	202.69241	5	217.93244
6	157.27711	6	172.51715	6	187.75718	6	202.99721	6	218.23724
7	157.58192	7	172.82195	7	188.06198	7	203.30201	7	218.54204
8	157.88672	8	173.12675	8	188.36678	8	203.60681	8	218.84684
9	158.19152	9	173.43155	9	188.67158	9	203.91161	9	219.15164
520	158.49632	570	173.73635	620	188.97638	670	204.21641	720	219.45644
1	158.80112	1	174.04115	1	189.28118	1	204.52121	1	219.76124
2	159.10592	2	174.34595	2	189.58598	2	204.82601	2	220.06604
3	159.41072	3	174.65075	3	189.89078	3	205.13081	3	220.37084
4	159.71552	4	174.95555	4	190.19558	4	205.43561	4	220.67564
5	160.02032	5	175.26035	5	190.50038	5	205.74041	5	220.98044
6	160.32512	6	175.56515	6	190.80518	6	206.04521	6	221.28524
7	160.62992	7	175.86995	7	191.10998	7	206.35001	7	221.59004
8	160.93472	8	176.17475	8	191.41478	8	206.65481	8	221.89484
9	161.23952	9	176.47955	9	191.71958	9	206.95961	9	222.19964
530	161.54432	580	176.78435	630	192.02438	680	207.26441	730	222.50445
1	161.84912	1	177.08915	1	192.32918	1	207.56922	1	222.80925
2	162.15392	2	177.39395	2	192.63399	2	207.87402	2	223.11405
3	162.45872	3	177.69876	3	192.93879	3	208.17882	3	223.41885
4	162.76353	4	178.00356	4	193.24359	4	208.48362	4	223.72365
5	163.06833	5	178.30836	5	193.54839	5	208.78842	5	224.02845
6	163.37313	6	178.61316	6	193.85319	6	209.09322	6	224.33325
7	163.67793	7	178.91796	7	194.15799	7	209.39802	7	224.63805
8	163.98273	8	179.22276	8	194.46279	8	209.70282	8	224.94285
9	164.28753	9	179.52756	9	194.76759	9	210.00762	9	225.24765
540	164.59233	590	179.83236	640	195.07239	690	210.31242	740	225.55245
1	164.89713	1	180.13716	1	195.37719	1	210.61722	1	225.85725
2	165.20193	2	180.44196	2	195.68199	2	210.92202	2	226.16205
3	165.50673	3	180.74676	3	195.98679	3	211.22682	3	226.46685
4	165.81153	4	181.05156	4	196.29159	4	211.53162	4	226.77165
5	166.11633	5	181.35636	5	196.59639	5	211.83642	5	227.07645
6	166.42113	6	181.66116	6	196.90119	6	212.14122	6	227.38125
7	166.72593	7	181.96596	7	197.20599	7	212.44602	7	227.68606
8	167.03073	8	182.27076	8	197.51080	8	212.75083	8	227.99086
9	167.33553	9	182.57557	9	197.81560	9	213.05563	9	228.29566

LENGTH—FEET TO METRES

[Reduction factor: 1 foot = 0.3048006096 metre]

Feet	Metres	Feet	Metres	Feet	Metres	Feet	Metres	Feet	Metres
750	228.60046	**800**	243.84049	**850**	259.08052	**900**	274.32055	**950**	289.56058
1	228.90526	1	244.14529	1	259.38532	1	274.62535	1	289.86538
2	229.21006	2	244.45009	2	259.69012	2	274.93015	2	290.17018
3	229.51486	3	244.75489	3	259.99492	3	275.23495	3	290.47498
4	229.81966	4	245.05969	4	260.29972	4	275.53975	4	290.77978
5	230.12446	5	245.36449	5	260.60452	5	275.84455	5	291.08458
6	230.42926	6	245.66929	6	260.90932	6	276.14935	6	291.38938
7	230.73406	7	245.97409	7	261.21412	7	276.45415	7	291.69418
8	231.03886	8	246.27889	8	261.51892	8	276.75895	8	291.99898
9	231.34366	9	246.58369	9	261.82372	9	277.06375	9	292.30378
760	231.64846	**810**	246.88849	**860**	262.12852	**910**	277.36855	**960**	292.60859
1	231.95326	1	247.19329	1	262.43332	1	277.67336	1	292.91339
2	232.25806	2	247.49809	2	262.73813	2	277.97816	2	293.21819
3	232.56287	3	247.80290	3	263.04293	3	278.28296	3	293.52299
4	232.86767	4	248.10770	4	263.34773	4	278.58776	4	293.82779
5	233.17247	5	248.41250	5	263.65253	5	278.89256	5	294.13259
6	233.47727	6	248.71730	6	263.95733	6	279.19736	6	294.43739
7	233.78207	7	249.02210	7	264.26213	7	279.50216	7	294.74219
8	234.08687	8	249.32690	8	264.56693	8	279.80696	8	295.04699
9	234.39167	9	249.63170	9	264.87173	9	280.11176	9	295.35179
770	234.69647	**820**	249.93650	**870**	265.17653	**920**	280.41656	**970**	295.65659
1	235.00127	1	250.24130	1	265.48133	1	280.72136	1	295.96139
2	235.30607	2	250.54610	2	265.78613	2	281.02616	2	296.26619
3	235.61087	3	250.85090	3	266.09093	3	281.33096	3	296.57099
4	235.91567	4	251.15570	4	266.39573	4	281.63576	4	296.87579
5	236.22047	5	251.46050	5	266.70053	5	281.94056	5	297.18059
6	236.52527	6	251.76530	6	267.00533	6	282.24536	6	297.48539
7	236.83007	7	252.07010	7	267.31013	7	282.55017	7	297.79020
8	237.13487	8	252.37490	8	267.61494	8	282.85497	8	298.09500
9	237.43967	9	252.67971	9	267.91974	9	283.15977	9	298.39980
780	237.74448	**830**	252.98451	**880**	268.22454	**930**	283.46457	**980**	298.70460
1	238.04928	1	253.28931	1	268.52934	1	283.76937	1	299.00940
2	238.35408	2	253.59411	2	268.83414	2	284.07417	2	299.31420
3	238.65888	3	253.89891	3	269.13894	3	284.37897	3	299.61900
4	238.96368	4	254.20371	4	269.44374	4	284.68377	4	299.92380
5	239.26848	5	254.50851	5	269.74854	5	284.98857	5	300.22860
6	239.57328	6	254.81331	6	270.05334	6	285.29337	6	300.53340
7	239.87808	7	255.11811	7	270.35814	7	285.59817	7	300.83820
8	240.18288	8	255.42291	8	270.66294	8	285.90297	8	301.14300
9	240.48768	9	255.72771	9	270.96774	9	286.20777	9	301.44780
790	240.79248	**840**	256.03251	**890**	271.27254	**940**	286.51257	**990**	301.75260
1	241.09728	1	256.33731	1	271.57734	1	286.81737	1	302.05740
2	241.40208	2	256.64211	2	271.88214	2	287.12217	2	302.36220
3	241.70688	3	256.94691	3	272.18694	3	287.42697	3	302.66701
4	242.01168	4	257.25171	4	272.49174	4	287.73178	4	302.97181
5	242.31648	5	257.55652	5	272.79655	5	288.03658	5	303.27661
6	242.62129	6	257.86132	6	273.10135	6	288.34138	6	303.58141
7	242.92609	7	258.16612	7	273.40615	7	288.64618	7	303.88621
8	243.23089	8	258.47092	8	273.71095	8	288.95098	8	304.19101
9	243.53569	9	258.77572	9	274.01575	9	289.25578	9	304.49581

LENGTH—KILOMETRES TO MILES

[Reduction factor: 1 kilometre = 0.6213699495 mile]

Kilo-metres	Miles	Kilo-metres	Miles	Kilo-metres	Miles	Kilo-metres	Miles	Kilo-metres	Miles
0		50	31.06850	100	62.13699	150	93.20549	200	124.27399
1	0.62137	1	31.68987	1	62.75836	1	93.82686	1	124.89536
2	1.24274	2	32.31124	2	63.37973	2	94.44823	2	125.51673
3	1.86411	3	32.93261	3	64.00110	3	95.06960	3	126.13810
4	2.48548	4	33.55398	4	64.62247	4	95.69097	4	126.75947
5	3.10685	5	34.17535	5	65.24384	5	96.31234	5	127.38084
6	3.72822	6	34.79672	6	65.86521	6	96.93371	6	128.00221
7	4.34959	7	35.41809	7	66.48658	7	97.55508	7	128.62358
8	4.97096	8	36.03946	8	67.10795	8	98.17645	8	129.24495
9	5.59233	9	36.66083	9	67.72932	9	98.79782	9	129.86632
10	6.21370	60	37.28220	110	68.35069	160	99.41919	210	130.48769
1	6.83507	1	37.90357	1	68.97206	1	100.04056	1	131.10906
2	7.45644	2	38.52494	2	69.59343	2	100.66193	2	131.73043
3	8.07781	3	39.14631	3	70.21480	3	101.28330	3	132.35180
4	8.69918	4	39.76768	4	70.83617	4	101.90467	4	132.97317
5	9.32055	5	40.38905	5	71.45754	5	102.52604	5	133.59454
6	9.94192	6	41.01042	6	72.07891	6	103.14741	6	134.21591
7	10.56329	7	41.63179	7	72.70028	7	103.76878	7	134.83728
8	11.18466	8	42.25316	8	73.32165	8	104.39015	8	135.45865
9	11.80603	9	42.87453	9	73.94302	9	105.01152	9	136.08002
20	12.42740	70	43.49590	120	74.56439	170	105.63289	220	136.70139
1	13.04877	1	44.11727	1	75.18576	1	106.25426	1	137.32276
2	13.67014	2	44.73864	2	75.80713	2	106.87563	2	137.94413
3	14.29151	3	45.36001	3	76.42850	3	107.49700	3	138.56550
4	14.91288	4	45.98138	4	77.04987	4	108.11837	4	139.18687
5	15.53425	5	46.60275	5	77.67124	5	108.73974	5	139.80824
6	16.15562	6	47.22412	6	78.29261	6	109.36111	6	140.42961
7	16.77699	7	47.84549	7	78.91398	7	109.98248	7	141.05098
8	17.39836	8	48.46686	8	79.53535	8	110.60385	8	141.67235
9	18.01973	9	49.08823	9	80.15672	9	111.22522	9	142.29372
30	18.64110	80	49.70960	130	80.77809	180	111.84659	230	142.91509
1	19.26247	1	50.33097	1	81.39946	1	112.46796	1	143.53646
2	19.88384	2	50.95234	2	82.02083	2	113.08933	2	144.15783
3	20.50521	3	51.57371	3	82.64220	3	113.71070	3	144.77920
4	21.12658	4	52.19508	4	83.26357	4	114.33207	4	145.40057
5	21.74795	5	52.81645	5	83.88494	5	114.95344	5	146.02194
6	22.36932	6	53.43782	6	84.50631	6	115.57481	6	146.64331
7	22.99069	7	54.05919	7	85.12768	7	116.19618	7	147.26468
8	23.61206	8	54.68056	8	85.74905	8	116.81755	8	147.88605
9	24.23343	9	55.30193	9	86.37042	9	117.43892	9	148.50742
40	24.85480	90	55.92330	140	86.99179	190	118.06029	240	149.12879
1	25.47617	1	56.54467	1	87.61316	1	118.68166	1	149.75016
2	26.09754	2	57.16604	2	88.23453	2	119.30303	2	150.37153
3	26.71091	3	57.78741	3	88.85590	3	119.92440	3	150.99290
4	27.34028	4	58.40878	4	89.47727	4	120.54577	4	151.61427
5	27.96165	5	59.03015	5	90.09864	5	121.16714	5	152.23564
6	28.58302	6	59.65152	6	90.72001	6	121.78851	6	152.85701
7	29.20439	7	60.27289	7	91.34138	7	122.40988	7	153.47838
8	29.82576	8	60.89426	8	91.96275	8	123.03125	8	154.09975
9	30.44713	9	61.51562	9	92.58412	9	123.65262	9	154.72112

LENGTH—KILOMETRES TO MILES

[Reduction factor: 1 kilometre = 0.6213699495 mile]

Kilo-metres	Miles	Kilo-metres	Miles	Kilo-metres	Miles	Kilo-metres	Miles	Kilo-metres	Miles
250	155.34249	300	186.41098	350	217.47948	400	248.54798	450	279.61648
1	155.96386	1	187.03235	1	218.10085	1	249.16935	1	280.23785
2	156.58523	2	187.65372	2	218.72222	2	249.79072	2	280.85922
3	157.20660	3	188.27509	3	219.34359	3	250.41209	3	281.48059
4	157.82797	4	188.89646	4	219.96496	4	251.03346	4	282.10196
5	158.44934	5	189.51783	5	220.58633	5	251.65483	5	282.72333
6	159.07071	6	190.13920	6	221.20770	6	252.27620	6	283.34470
7	159.69208	7	190.76057	7	221.82907	7	252.89757	7	283.96607
8	160.31345	8	191.38194	8	222.45044	8	253.51894	8	284.58744
9	160.93482	9	192.00331	9	223.07181	9	254.14031	9	285.20881
260	161.55619	310	192.62468	360	223.69318	410	254.76168	460	285.83018
1	162.17756	1	193.24605	1	224.31455	1	255.38305	1	286.45155
2	162.79893	2	193.86742	2	224.93592	2	256.00442	2	287.07292
3	163.42030	3	194.48879	3	225.55729	3	256.62579	3	287.69429
4	164.04167	4	195.11016	4	226.17866	4	257.24716	4	288.31566
5	164.66304	5	195.73153	5	226.80003	5	257.86853	5	288.93703
6	165.28441	6	196.35290	6	227.42140	6	258.48990	6	289.55840
7	165.90578	7	196.97427	7	228.04277	7	259.11127	7	290.17977
8	166.52715	8	197.59564	8	228.66414	8	259.73264	8	290.80114
9	167.14852	9	198.21701	9	229.28551	9	260.35401	9	291.42251
270	167.76989	320	198.83838	370	229.90688	420	260.97538	470	292.04388
1	168.39126	1	199.45975	1	230.52825	1	261.59675	1	292.66525
2	169.01263	2	200.08112	2	231.14962	2	262.21812	2	293.28662
3	169.63400	3	200.70249	3	231.77099	3	262.83949	3	293.90799
4	170.25537	4	201.32386	4	232.39236	4	263.46086	4	294.52936
5	170.87674	5	201.94523	5	233.01373	5	264.08223	5	295.15073
6	171.49811	6	202.56660	6	233.63510	6	264.70360	6	295.77210
7	172.11948	7	203.18797	7	234.25647	7	265.32497	7	296.39347
8	172.74085	8	203.80934	8	234.87784	8	265.94634	8	297.01484
9	173.36222	9	204.43071	9	235.49921	9	266.56771	9	297.63621
280	173.98359	330	205.05208	380	236.12058	430	267.18908	480	298.25758
1	174.60496	1	205.67345	1	236.74195	1	267.81045	1	298.87895
2	175.22633	2	206.29482	2	237.36332	2	268.43182	2	299.50032
3	175.84770	3	206.91619	3	237.98469	3	269.05319	3	300.12169
4	176.46907	4	207.53756	4	238.60606	4	269.67456	4	300.74306
5	177.09044	5	208.15893	5	239.22743	5	270.29593	5	301.36443
6	177.71181	6	208.78030	6	239.84880	6	270.91730	6	301.98580
7	178.33318	7	209.40167	7	240.47017	7	271.53867	7	302.60717
8	178.95455	8	210.02304	8	241.09154	8	272.16004	8	303.22854
9	179.57592	9	210.64441	9	241.71291	9	272.78141	9	303.84991
290	180.19729	340	211.26578	390	242.33428	440	273.40278	490	304.47128
1	180.81866	1	211.88715	1	242.95565	1	274.02415	1	305.09265
2	181.44003	2	212.50852	2	243.57702	2	274.64552	2	305.71402
3	182.06140	3	213.12989	3	244.19839	3	275.26689	3	306.33539
4	182.68277	4	213.75126	4	244.81976	4	275.88826	4	306.95676
5	183.30414	5	214.37263	5	245.44113	5	276.50963	5	307.57812
6	183.92551	6	214.99400	6	246.06250	6	277.13100	6	308.19949
7	184.54687	7	215.61537	7	246.68387	7	277.75237	7	308.82086
8	185.16824	8	216.23674	8	247.30524	8	278.37374	8	309.44223
9	185.78961	9	216.85811	9	247.92661	9	278.99511	9	310.06360

LENGTH—KILOMETRES TO MILES

[Reduction factor: 1 kilometre = 0.6213699495 mile]

Kilo-metres	Miles	Kilo-metres	Miles	Kilo-metres	Miles	Kilo-metres	Miles	Kilo-metres	Miles
500	310.68497	550	341.75347	600	372.82197	650	403.89047	700	434.95896
1	311.30634	1	342.37484	1	373.44334	1	404.51184	1	435.58033
2	311.92771	2	342.99621	2	374.06471	2	405.13321	2	436.20170
3	312.54908	3	343.61758	3	374.68608	3	405.75458	3	436.82307
4	313.17045	4	344.23895	4	375.30745	4	406.37595	4	437.44444
5	313.79182	5	344.86032	5	375.92882	5	406.99732	5	438.06581
6	314.41319	6	345.48169	6	376.55019	6	407.61869	6	438.68718
7	315.03456	7	346.10306	7	377.17156	7	408.24006	7	439.30855
8	315.65593	8	346.72443	8	377.79293	8	408.86143	8	439.92992
9	316.27730	9	347.34580	9	378.41430	9	409.48280	9	440.55129
510	316.89867	560	347.96717	610	379.03567	660	410.10417	710	441.17266
1	317.52004	1	348.58854	1	379.65704	1	410.72554	1	441.79403
2	318.14141	2	349.20991	2	380.27841	2	411.34691	2	442.41540
3	318.76278	3	349.83128	3	380.89978	3	411.96828	3	443.03677
4	319.38415	4	350.45265	4	381.52115	4	412.58965	4	443.65814
5	320.00552	5	351.07402	5	382.14252	5	413.21102	5	444.27951
6	320.62689	6	351.69539	6	382.76389	6	413.83239	6	444.90088
7	321.24826	7	352.31676	7	383.38526	7	414.45376	7	445.52225
8	321.86963	8	352.93813	8	384.00663	8	415.07513	8	446.14362
9	322.49100	9	353.55950	9	384.62800	9	415.69650	9	446.76499
520	323.11237	570	354.18087	620	385.24937	670	416.31787	720	447.38636
1	323.73374	1	354.80224	1	385.87074	1	416.93924	1	448.00773
2	324.35511	2	355.42361	2	386.49211	2	417.56061	2	448.62910
3	324.97648	3	356.04498	3	387.11348	3	418.18198	3	449.25047
4	325.59785	4	356.66635	4	387.73485	4	418.80335	4	449.87184
5	326.21922	5	357.28772	5	388.35622	5	419.42472	5	450.49321
6	326.84059	6	357.90909	6	388.97759	6	420.04609	6	451.11458
7	327.46196	7	358.53046	7	389.59896	7	420.66746	7	451.73595
8	328.08333	8	359.15183	8	390.22033	8	421.28883	8	452.35732
9	328.70470	9	359.77320	9	390.84170	9	421.91020	9	452.97869
530	329.32607	580	360.39457	630	391.46307	680	422.53157	730	453.60006
1	329.94744	1	361.01594	1	392.08444	1	423.15294	1	454.22143
2	330.56881	2	361.63731	2	392.70581	2	423.77431	2	454.84280
3	331.19018	3	362.25868	3	393.32718	3	424.39568	3	455.46417
4	331.81155	4	362.88005	4	393.94855	4	425.01705	4	456.08554
5	332.43292	5	363.50142	5	394.56992	5	425.63842	5	456.70691
6	333.05429	6	364.12279	6	395.19129	6	426.25979	6	457.32828
7	333.67566	7	364.74416	7	395.81266	7	426.88116	7	457.94965
8	334.29703	8	365.36553	8	396.43403	8	427.50253	8	458.57102
9	334.91840	9	365.98690	9	397.05540	9	428.12390	9	459.19239
540	335.53977	590	366.60827	640	397.67677	690	428.74527	740	459.81376
1	336.16114	1	367.22964	1	398.29814	1	429.36664	1	460.43513
2	336.78251	2	367.85101	2	398.91951	2	429.98801	2	461.05650
3	337.40388	3	368.47238	3	399.54088	3	430.60937	3	461.67787
4	338.02525	4	369.09375	4	400.16225	4	431.23074	4	462.29924
5	338.64662	5	369.71512	5	400.78362	5	431.85211	5	462.92061
6	339.26799	6	370.33649	6	401.40499	6	432.47348	6	463.54198
7	339.88936	7	370.95786	7	402.02636	7	433.09485	7	464.16335
8	340.51073	8	371.57923	8	402.64773	8	433.71622	8	464.78472
9	341.13210	9	372.20060	9	403.26910	9	434.33759	9	465.40609

LENGTH—KILOMETRES TO MILES

[Reduction factor: 1 kilometre = 0.6213699495 mile]

Kilo-metres	Miles	Kilo-metres	Miles	Kilo-metres	Miles	Kilo-metres	Miles	Kilo-metres	Miles
750	466.02746	800	497.09596	850	528.16446	900	559.23295	950	590.30145
1	466.64883	1	497.71733	1	528.78583	1	559.85432	1	590.92282
2	467.27020	2	498.33870	2	529.40720	2	560.47569	2	591.54419
3	467.89157	3	498.96007	3	530.02857	3	561.09706	3	592.16556
4	468.51294	4	499.58144	4	530.64994	4	561.71843	4	592.78693
5	469.13431	5	500.20281	5	531.27131	5	562.33980	5	593.40830
6	469.75568	6	500.82418	6	531.89268	6	562.96117	6	594.02967
7	470.37705	7	501.44555	7	532.51405	7	563.58254	7	594.65104
8	470.99842	8	502.06692	8	533.13542	8	564.20391	8	595.27241
9	471.61979	9	502.68829	9	533.75679	9	564.82528	9	595.89378
760	472.24116	810	503.30966	860	534.37816	910	565.44665	960	596.51515
1	472.86253	1	503.93103	1	534.99953	1	566.06802	1	597.13652
2	473.48390	2	504.55240	2	535.62090	2	566.68939	2	597.75789
3	474.10527	3	505.17377	3	536.24227	3	567.31076	3	598.37926
4	474.72664	4	505.79514	4	536.86364	4	567.93213	4	599.00063
5	475.34801	5	506.41651	5	537.48501	5	568.55350	5	599.62200
6	475.96938	6	507.03788	6	538.10638	6	569.17487	6	600.24337
7	476.59075	7	507.65925	7	538.72775	7	569.79624	7	600.86474
8	477.21212	8	508.28062	8	539.34912	8	570.41761	8	601.48611
9	477.83349	9	508.90199	9	539.97049	9	571.03898	9	602.10748
770	478.45486	820	509.52336	870	540.59186	920	571.66035	970	602.72885
1	479.07623	1	510.14473	1	541.21323	1	572.28172	1	603.35022
2	479.69760	2	510.76610	2	541.83460	2	572.90309	2	603.97159
3	480.31897	3	511.38747	3	542.45597	3	573.52446	3	604.59296
4	480.94034	4	512.00884	4	543.07734	4	574.14583	4	605.21433
5	481.56171	5	512.63021	5	543.69871	5	574.76720	5	605.83570
6	482.18308	6	513.25158	6	544.32008	6	575.38857	6	606.45707
7	482.80445	7	513.87295	7	544.94145	7	576.00994	7	607.07844
8	483.42582	8	514.49432	8	545.56282	8	576.63131	8	607.69981
9	484.04719	9	515.11569	9	546.18419	9	577.25268	9	608.32118
780	484.66856	830	515.73706	880	546.80556	930	577.87405	980	608.94255
1	485.28993	1	516.35843	1	547.42693	1	578.49542	1	609.56392
2	485.91130	2	516.97980	2	548.04830	2	579.11679	2	610.18529
3	486.53267	3	517.60117	3	548.66967	3	579.73816	3	610.80666
4	487.15404	4	518.22254	4	549.29104	4	580.35953	4	611.42803
5	487.77541	5	518.84391	5	549.91241	5	580.98090	5	612.04940
6	488.39678	6	519.46528	6	550.53378	6	581.60227	6	612.67077
7	489.01815	7	520.08665	7	551.15515	7	582.22364	7	613.29214
8	489.63952	8	520.70802	8	551.77652	8	582.84501	8	613.91351
9	490.26089	9	521.32939	9	552.39789	9	583.46638	9	614.53488
790	490.88226	840	521.95076	890	553.01926	940	584.08775	990	615.15625
1	491.50363	1	522.57213	1	553.64062	1	584.70912	1	615.77762
2	492.12500	2	523.19350	2	554.26199	2	585.33049	2	616.39899
3	492.74637	3	523.81487	3	554.88336	3	585.95186	3	617.02036
4	493.36774	4	524.43624	4	555.50473	4	586.57323	4	617.64173
5	493.98911	5	525.05761	5	556.12610	5	587.19460	5	618.26310
6	494.61048	6	525.67898	6	556.74747	6	587.81597	6	618.88447
7	495.23185	7	526.30035	7	557.36884	7	588.43734	7	619.50584
8	495.85322	8	526.92172	8	557.99021	8	589.05871	8	620.12721
9	496.47459	9	527.54309	9	558.61158	9	589.68008	9	620.74858

LENGTH—MILES TO KILOMETRES

[Reduction factor: 1 mile = 1.609347219 kilometres]

Miles	Kilo-metres	Miles	Kilo-metres	Miles	Kilo-metres	Miles	Kilo-metres	Miles	Kilo-metres
0		50	80.4674	100	160.9347	150	241.4021	200	321.8694
1	1.6093	1	82.0767	1	162.5441	1	243.0114	1	323.4788
2	3.2187	2	83.6861	2	164.1534	2	244.6208	2	325.0881
3	4.8280	3	85.2954	3	165.7628	3	246.2301	3	326.6975
4	6.4374	4	86.9047	4	167.3721	4	247.8395	4	328.3068
5	8.0467	5	88.5141	5	168.9815	5	249.4488	5	329.9162
6	9.6561	6	90.1234	6	170.5908	6	251.0582	6	331.5255
7	11.2654	7	91.7328	7	172.2002	7	252.6675	7	333.1349
8	12.8748	8	93.3421	8	173.8095	8	254.2769	8	334.7442
9	14.4841	9	94.9515	9	175.4188	9	255.8862	9	336.3536
10	16.0935	60	96.5608	110	177.0282	160	257.4956	210	337.9629
1	17.7028	1	98.1702	1	178.6375	1	259.1049	1	339.5723
2	19.3122	2	99.7795	2	180.2469	2	260.7142	2	341.1816
3	20.9215	3	101.3889	3	181.8562	3	262.3236	3	342.7910
4	22.5309	4	102.9982	4	183.4656	4	263.9329	4	344.4003
5	24.1402	5	104.6076	5	185.0749	5	265.5423	5	346.0097
6	25.7496	6	106.2169	6	186.6843	6	267.1516	6	347.6190
7	27.3589	7	107.8263	7	188.2936	7	268.7610	7	349.2283
8	28.9682	8	109.4356	8	189.9030	8	270.3703	8	350.8377
9	30.5776	9	111.0450	9	191.5123	9	271.9797	9	352.4470
20	32.1869	70	112.6543	120	193.1217	170	273.5890	220	354.0564
1	33.7963	1	114.2637	1	194.7310	1	275.1984	1	355.6657
2	35.4056	2	115.8730	2	196.3404	2	276.8077	2	357.2751
3	37.0150	3	117.4823	3	197.9497	3	278.4171	3	358.8844
4	38.6243	4	119.0917	4	199.5591	4	280.0264	4	360.4938
5	40.2337	5	120.7010	5	201.1684	5	281.6358	5	362.1031
6	41.8430	6	122.3104	6	202.7777	6	283.2451	6	363.7125
7	43.4524	7	123.9197	7	204.3871	7	284.8545	7	365.3218
8	45.0617	8	125.5291	8	205.9954	8	286.4638	8	366.9312
9	46.6711	9	127.1384	9	207.6058	9	288.0732	9	368.5405
30	48.2804	80	128.7478	130	209.2151	180	289.6825	230	370.1499
1	49.8898	1	130.3571	1	210.8245	1	291.2918	1	371.7592
2	51.4991	2	131.9665	2	212.4338	2	292.9012	2	373.3686
3	53.1085	3	133.5758	3	214.0432	3	294.5105	3	374.9779
4	54.7178	4	135.1852	4	215.6525	4	296.1199	4	376.5872
5	56.3272	5	136.7945	5	217.2619	5	297.7292	5	378.1966
6	57.9365	6	138.4039	6	218.8712	6	299.3386	6	379.8059
7	59.5458	7	140.0132	7	220.4806	7	300.9479	7	381.4153
8	61.1552	8	141.6226	8	222.0899	8	302.5573	8	383.0246
9	62.7645	9	143.2319	9	223.6993	9	304.1666	9	384.6340
40	64.3739	90	144.8412	140	225.3086	190	305.7760	240	386.2433
1	65.9832	1	146.4506	1	226.9180	1	307.3853	1	387.8527
2	67.5926	2	148.0599	2	228.5273	2	308.9947	2	389.4620
3	69.2019	3	149.6693	3	230.1366	3	310.6040	3	391.0714
4	70.8113	4	151.2786	4	231.7460	4	312.2134	4	392.6807
5	72.4206	5	152.8880	5	233.3553	5	313.8227	5	394.2901
6	74.0300	6	154.4973	6	234.9647	6	315.4321	6	395.8994
7	75.6393	7	156.1067	7	236.5740	7	317.0414	7	397.5088
8	77.2487	8	157.7160	8	238.1834	8	318.6507	8	399.1181
9	78.8580	9	159.3254	9	239.7927	9	320.2601	9	400.7275

LENGTH—MILES TO KILOMETRES

[Reduction factor: 1 mile = 1.609347219 kilometres]

Miles	Kilo-metres	Miles	Kilo-metres	Miles	Kilo-metres	Miles	Kilo-metres	Miles	Kilo-metres
250	402.3368	300	482.8042	350	563.2715	400	643.7389	450	724.2062
1	403.9461	1	484.4135	1	564.8809	1	645.3482	1	725.8156
2	405.5555	2	486.0229	2	566.4902	2	646.9576	2	727.4249
3	407.1648	3	487.6322	3	568.0996	3	648.5669	3	729.0343
4	408.7742	4	489.2416	4	569.7089	4	650.1763	4	730.6436
5	410.3835	5	490.8509	5	571.3183	5	651.7856	5	732.2530
6	411.9929	6	492.4602	6	572.9276	6	653.3950	6	733.8623
7	413.6022	7	494.0696	7	574.5370	7	655.0043	7	735.4717
8	415.2116	8	495.6789	8	576.1463	8	656.6137	8	737.0810
9	416.8209	9	497.2883	9	577.7557	9	658.2230	9	738.6904
260	418.4303	310	498.8976	360	579.3650	410	659.8824	460	740.2997
1	420.0396	1	500.5070	1	580.9743	1	661.4417	1	741.9091
2	421.6490	2	502.1163	2	582.5837	2	663.0511	2	743.5184
3	423.2583	3	503.7257	3	584.1930	3	664.6604	3	745.1278
4	424.8677	4	505.3350	4	585.8024	4	666.2697	4	746.7371
5	426.4770	5	506.9444	5	587.4117	5	667.8791	5	748.3465
6	428.0864	6	508.5537	6	589.0211	6	669.4884	6	749.9558
7	429.6957	7	510.1631	7	590.6304	7	671.0978	7	751.5652
8	431.3051	8	511.7724	8	592.2398	8	672.7071	8	753.1745
9	432.9144	9	513.3818	9	593.8491	9	674.3165	9	754.7838
270	434.5237	320	514.9911	370	595.4585	420	675.9258	470	756.3932
1	436.1331	1	516.6005	1	597.0678	1	677.5352	1	758.0025
2	437.7424	2	518.2098	2	598.6772	2	679.1445	2	759.6119
3	439.3518	3	519.8192	3	600.2865	3	680.7539	3	761.2212
4	440.9611	4	521.4285	4	601.8959	4	682.3632	4	762.8306
5	442.5705	5	523.0378	5	603.5052	5	683.9726	5	764.4399
6	444.1798	6	524.6472	6	605.1145	6	685.5819	6	766.0493
7	445.7892	7	526.2565	7	606.7239	7	687.1913	7	767.6586
8	447.3985	8	527.8659	8	608.3332	8	688.8006	8	769.2680
9	449.0079	9	529.4752	9	609.9426	9	690.4100	9	770.8773
280	450.6172	330	531.0846	380	611.5519	430	692.0193	480	772.4867
1	452.2266	1	532.6939	1	613.1613	1	693.6287	1	774.0960
2	453.8359	2	534.3033	2	614.7706	2	695.2380	2	775.7054
3	455.4453	3	535.9126	3	616.3800	3	696.8473	3	777.3147
4	457.0546	4	537.5220	4	617.9893	4	698.4567	4	778.9241
5	458.6640	5	539.1313	5	619.5987	5	700.0660	5	780.5334
6	460.2733	6	540.7407	6	621.2080	6	701.6754	6	782.1427
7	461.8827	7	542.3500	7	622.8174	7	703.2847	7	783.7521
8	463.4920	8	543.9594	8	624.4267	8	704.8941	8	785.3614
9	465.1013	9	545.5687	9	626.0361	9	706.5034	9	786.9708
290	466.7107	340	547.1781	390	627.6454	440	708.1128	490	788.5801
1	468.3200	1	548.7874	1	629.2548	1	709.7221	1	790.1895
2	469.9294	2	550.3967	2	630.8641	2	711.3315	2	791.7988
3	471.5387	3	552.0061	3	632.4735	3	712.9408	3	793.4082
4	473.1481	4	553.6154	4	634.0828	4	714.5502	4	795.0175
5	474.7574	5	555.2248	5	635.6922	5	716.1595	5	796.6269
6	476.3668	6	556.8341	6	637.3015	6	717.7689	6	798.2362
7	477.9761	7	558.4435	7	638.9108	7	719.3782	7	799.8456
8	479.5855	8	560.0528	8	640.5202	8	720.9876	8	801.4549
9	481.1948	9	561.6622	9	642.1295	9	722.5969	9	803.0643

LENGTH—MILES TO KILOMETRES

[Reduction factor: 1 mile = 1.609347219 kilometres]

Miles	Kilo-metres	Miles	Kilo-metres	Miles	Kilo-metres	Miles	Kilo-metres	Miles	Kilo-metres
500	804.6736	550	885.1410	600	965.6083	650	1,046.0757	700	1,126.5431
1	806.2830	1	886.7503	1	967.2177	1	1,047.6850	1	1,128.1524
2	807.8923	2	888.3597	2	968.8270	2	1,049.2944	2	1,129.7617
3	809.5017	3	889.9690	3	970.4364	3	1,050.9037	3	1,131.3711
4	811.1110	4	891.5784	4	972.0457	4	1,052.5131	4	1,132.9804
5	812.7203	5	893.1877	5	973.6551	5	1,054.1224	5	1,134.5898
6	814.3297	6	894.7971	6	975.2644	6	1,055.7318	6	1,136.1991
7	815.9390	7	896.4064	7	976.8738	7	1,057.3411	7	1,137.8085
8	817.5484	8	898.0157	8	978.4831	8	1,058.9505	8	1,139.4178
9	819.1577	9	899.6251	9	980.0925	9	1,060.5598	9	1,141.0272
510	820.7671	560	901.2344	610	981.7018	660	1,062.1692	710	1,142.6365
1	822.3764	1	902.8438	1	983.3112	1	1,063.7785	1	1,144.2459
2	823.9858	2	904.4531	2	984.9205	2	1,065.3879	2	1,145.8552
3	825.5951	3	906.0625	3	986.5298	3	1,066.9972	3	1,147.4646
4	827.2045	4	907.6718	4	988.1392	4	1,068.6066	4	1,149.0739
5	828.8138	5	909.2812	5	989.7485	5	1,070.2159	5	1,150.6833
6	830.4232	6	910.8905	6	991.3579	6	1,071.8252	6	1,152.2926
7	832.0325	7	912.4999	7	992.9672	7	1,073.4346	7	1,153.9020
8	833.6419	8	914.1092	8	994.5766	8	1,075.0439	8	1,155.5113
9	835.2512	9	915.7186	9	996.1859	9	1,076.6533	9	1,157.1207
520	836.8606	570	917.3279	620	997.7953	670	1,078.2626	720	1,158.7300
1	838.4699	1	918.9373	1	999.4046	1	1,079.8720	1	1,160.3393
2	840.0792	2	920.5466	2	1,001.0140	2	1,081.4813	2	1,161.9487
3	841.6886	3	922.1560	3	1,002.6233	3	1,083.0907	3	1,163.5580
4	843.2979	4	923.7653	4	1,004.2327	4	1,084.7000	4	1,165.1674
5	844.9073	5	925.3747	5	1,005.8420	5	1,086.3094	5	1,166.7767
6	846.5166	6	926.9840	6	1,007.4514	6	1,087.9187	6	1,168.3861
7	848.1260	7	928.5933	7	1,009.0607	7	1,089.5281	7	1,169.9954
8	849.7353	8	930.2027	8	1,010.6701	8	1,091.1374	8	1,171.6048
9	851.3447	9	931.8120	9	1,012.2794	9	1,092.7468	9	1,173.2141
530	852.9540	580	933.4214	630	1,013.8887	680	1,094.3561	730	1,174.8235
1	854.5634	1	935.0307	1	1,015.4981	1	1,095.9655	1	1,176.4328
2	856.1727	2	936.6401	2	1,017.1074	2	1,097.5748	2	1,178.0422
3	857.7821	3	938.2494	3	1,018.7168	3	1,099.1842	3	1,179.6515
4	859.3914	4	939.8588	4	1,020.3261	4	1,100.7935	4	1,181.2609
5	861.0008	5	941.4681	5	1,021.9355	5	1,102.4028	5	1,182.8702
6	862.6101	6	943.0775	6	1,023.5448	6	1,104.0122	6	1,184.4796
7	864.2195	7	944.6868	7	1,025.1542	7	1,105.6215	7	1,186.0889
8	865.8288	8	946.2962	8	1,026.7635	8	1,107.2309	8	1,187.6982
9	867.4382	9	947.9055	9	1,028.3729	9	1,108.8402	9	1,189.3076
540	869.0475	590	949.5149	640	1,029.9822	690	1,110.4496	740	1,190.9169
1	870.6568	1	951.1242	1	1,031.5916	1	1,112.0589	1	1,192.5263
2	872.2662	2	952.7336	2	1,033.2009	2	1,113.6683	2	1,194.1356
3	873.8755	3	954.3429	3	1,034.8103	3	1,115.2776	3	1,195.7450
4	875.4849	4	955.9522	4	1,036.4196	4	1,116.8870	4	1,197.3543
5	877.0942	5	957.5616	5	1,038.0290	5	1,118.4963	5	1,198.9637
6	878.7036	6	959.1709	6	1,039.6383	6	1,120.1057	6	1,200.5730
7	880.3129	7	960.7803	7	1,041.2477	7	1,121.7150	7	1,202.1824
8	881.9223	8	962.3896	8	1,042.8570	8	1,123.3244	8	1,203.7917
9	883.5316	9	963.9990	9	1,044.4663	9	1,124.9337	9	1,205.4011

LENGTH—MILES TO KILOMETRES

[Reduction factor: 1 mile = 1.609347219 kilometres]

Miles	Kilo-metres	Miles	Kilo-metres	Miles	Kilo-metres	Miles	Kilo-metres	Miles	Kilo-metres
750	1,207.0104	800	1,287.4778	850	1,367.9451	900	1,448.4125	950	1,528.8799
1	1,208.6198	1	1,289.0871	1	1,369.5545	1	1,450.0218	1	1,530.4892
2	1,210.2291	2	1,290.6965	2	1,371.1638	2	1,451.6312	2	1,532.0986
3	1,211.8385	3	1,292.3058	3	1,372.7732	3	1,453.2405	3	1,533.7079
4	1,213.4478	4	1,293.9152	4	1,374.3825	4	1,454.8499	4	1,535.3172
5	1,215.0572	5	1,295.5245	5	1,375.9919	5	1,456.4592	5	1,536.9266
6	1,216.6665	6	1,297.1339	6	1,377.6012	6	1,458.0686	6	1,538.5359
7	1,218.2758	7	1,298.7432	7	1,379.2106	7	1,459.6779	7	1,540.1453
8	1,219.8852	8	1,300.3526	8	1,380.8199	8	1,461.2873	8	1,541.7546
9	1,221.4945	9	1,301.9619	9	1,382.4293	9	1,462.8966	9	1,543.3640
760	1,223.1039	810	1,303.5712	860	1,384.0386	910	1,464.5060	960	1,544.9733
1	1,224.7132	1	1,305.1806	1	1,385.6480	1	1,466.1153	1	1,546.5827
2	1,226.3226	2	1,306.7899	2	1,387.2573	2	1,467.7247	2	1,548.1920
3	1,227.9319	3	1,308.3993	3	1,388.8666	3	1,469.3340	3	1,549.8014
4	1,229.5413	4	1,310.0086	4	1,390.4760	4	1,470.9434	4	1,551.4107
5	1,231.1506	5	1,311.6180	5	1,392.0853	5	1,472.5527	5	1,553.0201
6	1,232.7600	6	1,313.2273	6	1,393.6947	6	1,474.1621	6	1,554.6294
7	1,234.3693	7	1,314.8367	7	1,395.3040	7	1,475.7714	7	1,556.2388
8	1,235.9787	8	1,316.4460	8	1,396.9134	8	1,477.3807	8	1,557.8481
9	1,237.5880	9	1,318.0554	9	1,398.5227	9	1,478.9901	9	1,559.4575
770	1,239.1974	820	1,319.6647	870	1,400.1321	920	1,480.5994	970	1,561.0668
1	1,240.8067	1	1,321.2741	1	1,401.7414	1	1,482.2088	1	1,562.6761
2	1,242.4161	2	1,322.8834	2	1,403.3508	2	1,483.8181	2	1,564.2855
3	1,244.0254	3	1,324.4928	3	1,404.9601	3	1,485.4275	3	1,565.8948
4	1,245.6347	4	1,326.1021	4	1,406.5695	4	1,487.0368	4	1,567.5042
5	1,247.2441	5	1,327.7115	5	1,408.1788	5	1,488.6462	5	1,569.1135
6	1,248.8534	6	1,329.3208	6	1,409.7882	6	1,490.2555	6	1,570.7229
7	1,250.4628	7	1,330.9301	7	1,411.3975	7	1,491.8649	7	1,572.3322
8	1,252.0721	8	1,332.5395	8	1,413.0069	8	1,493.4742	8	1,573.9416
9	1,253.6815	9	1,334.1488	9	1,414.6162	9	1,495.0836	9	1,575.5509
780	1,255.2908	830	1,335.7582	880	1,416.2256	930	1,496.6929	980	1,577.1603
1	1,256.9002	1	1,337.3675	1	1,417.8349	1	1,498.3023	1	1,578.7696
2	1,258.5095	2	1,338.9769	2	1,419.4442	2	1,499.9116	2	1,580.3790
3	1,260.1189	3	1,340.5862	3	1,421.0536	3	1,501.5210	3	1,581.9883
4	1,261.7282	4	1,342.1956	4	1,422.6629	4	1,503.1303	4	1,583.5977
5	1,263.3376	5	1,343.8049	5	1,424.2723	5	1,504.7396	5	1,585.2070
6	1,264.9469	6	1,345.4143	6	1,425.8816	6	1,506.3490	6	1,586.8164
7	1,266.5563	7	1,347.0236	7	1,427.4910	7	1,507.9583	7	1,588.4257
8	1,268.1656	8	1,348.6330	8	1,429.1003	8	1,509.5677	8	1,590.0351
9	1,269.7750	9	1,350.2423	9	1,430.7097	9	1,511.1770	9	1,591.6444
790	1,271.3843	840	1,351.8517	890	1,432.3190	940	1,512.7864	990	1,593.2537
1	1,272.9936	1	1,353.4610	1	1,433.9284	1	1,514.3957	1	1,594.8631
2	1,274.6030	2	1,355.0704	2	1,435.5377	2	1,516.0051	2	1,596.4724
3	1,276.2123	3	1,356.6797	3	1,437.1471	3	1,517.6144	3	1 598.0818
4	1,277.8217	4	1,358.2891	4	1,438.7564	4	1,519.2238	4	1,599.6911
5	1,279.4310	5	1,359.8984	5	1,440.3658	5	1,520.8331	5	1,601.3005
6	1,281.0404	6	1,361.5077	6	1,441.9751	6	1,522.4425	6	1,602.9098
7	1,282.6497	7	1,363.1171	7	1,443.5845	7	1,524.0518	7	1,604.5192
8	1,284.2591	8	1,364.7264	8	1,445.1938	8	1,525.6612	8	1,606.1285
9	1,285.8684	9	1,366.3358	9	1,446.8031	9	1,527.2705	9	1,607.7379

AREA—HECTARES TO ACRES

[Reduction factor: 1 hectare = 2.471043930 acres]

Hectares	Acres	Hectares	Acres	Hectares	Acres	Hectares	Acres	Hectares	Acres
0		50	123.55220	100	247.10439	150	370.65659	200	494.20879
1	2.47104	1	126.02324	1	249.57544	1	373.12763	1	496.67983
2	4.94209	2	128.49428	2	252.04648	2	375.59888	2	499.15087
3	7.41313	3	130.96533	3	254.51752	3	378.06972	3	501.62192
4	9.88418	4	133.43637	4	256.98857	4	380.54077	4	504.09296
5	12.35522	5	135.90742	5	259.45961	5	383.01181	5	506.56401
6	14.82626	6	138.37846	6	261.93066	6	385.48285	6	509.03505
7	17.29731	7	140.84950	7	264.40170	7	387.95390	7	511.50609
8	19.76835	8	143.32055	8	266.87274	8	390.42494	8	513.97714
9	22.23940	9	145.79159	9	269.34379	9	392.89598	9	516.44818
10	24.71044	60	148.26264	110	271.81483	160	395.36703	210	518.91923
1	27.18148	1	150.73368	1	274.28588	1	397.83807	1	521.39027
2	29.65253	2	153.20472	2	276.75692	2	400.30912	2	523.86131
3	32.12357	3	155.67577	3	279.22796	3	402.78016	3	526.33236
4	34.59462	4	158.14681	4	281.69901	4	405.25120	4	528.80340
5	37.06566	5	160.61786	5	284.17005	5	407.72225	5	531.27444
6	39.53670	6	163.08890	6	286.64110	6	410.19329	6	533.74549
7	42.00775	7	165.55994	7	289.11214	7	412.66434	7	536.21653
8	44.47879	8	168.03099	8	291.58318	8	415.13538	8	538.68758
9	46.94983	9	170.50203	9	294.05423	9	417.60642	9	541.15862
20	49.42088	70	172.97308	120	296.52527	170	420.07747	220	543.62966
1	51.89192	1	175.44412	1	298.99632	1	422.54851	1	546.10071
2	54.36297	2	177.91516	2	301.46736	2	425.01956	2	548.57175
3	56.83401	3	180.38621	3	303.93840	3	427.49060	3	551.04280
4	59.30505	4	182.85725	4	306.40945	4	429.96164	4	553.51384
5	61.77610	5	185.32829	5	308.88049	5	432.43269	5	555.98488
6	64.24714	6	187.79934	6	311.35154	6	434.90373	6	558.45593
7	66.71819	7	190.27038	7	313.82258	7	437.37478	7	560.92697
8	69.18923	8	192.74143	8	316.29362	8	439.84582	8	563.39802
9	71.66027	9	195.21247	9	318.76467	9	442.31686	9	565.86906
30	74.13132	80	197.68351	130	321.23571	180	444.78791	230	568.34010
1	76.60236	1	200.15456	1	323.70675	1	447.25895	1	570.81115
2	79.07341	2	202.62560	2	326.17780	2	449.73000	2	573.28219
3	81.54445	3	205.09665	3	328.64884	3	452.20104	3	575.75324
4	84.01549	4	207.56769	4	331.11989	4	454.67208	4	578.22428
5	86.48654	5	210.03873	5	333.59093	5	457.14313	5	580.69532
6	88.95758	6	212.50978	6	336.06197	6	459.61417	6	583.16637
7	91.42863	7	214.98082	7	338.53302	7	462.08521	7	585.63741
8	93.89967	8	217.45187	8	341.00406	8	464.55626	8	588.10846
9	96.37071	9	219.92291	9	343.47511	9	467.02730	9	590.57950
40	98.84176	90	222.39395	140	345.94615	190	469.49835	240	593.05054
1	101.31280	1	224.86500	1	348.41719	1	471.96939	1	595.52159
2	103.78385	2	227.33604	2	350.88824	2	474.44043	2	597.99263
3	106.25489	3	229.80709	3	353.35928	3	476.91148	3	600.46367
4	108.72593	4	232.27813	4	355.83033	4	479.38252	4	602.93472
5	111.19698	5	234.74917	5	358.30137	5	481.85357	5	605.40576
6	113.66802	6	237.22022	6	360.77241	6	484.32461	6	607.87681
7	116.13906	7	239.69126	7	363.24346	7	486.79565	7	610.34785
8	118.61011	8	242.16231	8	365.71450	8	489.26670	8	612.81889
9	121.08115	9	244.63335	9	368.18555	9	491.73774	9	615.28994

AREA—HECTARES TO ACRES

[Reduction factor: 1 hectare = 2.471043930 acres]

Hectares	Acres	Hectares	Acres	Hectares	Acres	Hectares	Acres	Hectares	Acres
250	617.76098	300	741.31318	350	864.86538	400	988.41757	450	1,111.96977
1	620.23203	1	743.78422	1	867.33642	1	990.88862	1	1,114.44081
2	622.70307	2	746.25527	2	869.80746	2	993.35966	2	1,116.91186
3	625.17411	3	748.72631	3	872.27851	3	995.83070	3	1,119.38290
4	627.64516	4	751.19735	4	874.74955	4	998.30175	4	1,121.85394
5	630.11620	5	753.66840	5	877.22060	5	1,000.77279	5	1,124.32499
6	632.58725	6	756.13944	6	879.69164	6	1,003.24384	6	1,126.79603
7	635.05829	7	758.61049	7	882.16268	7	1,005.71488	7	1,129.26708
8	637.52933	8	761.08153	8	884.63373	8	1,008.18592	8	1,131.73812
9	640.00038	9	763.55257	9	887.10477	9	1,010.65697	9	1,134.20916
260	642.47142	310	766.02362	360	889.57581	410	1,013.12801	460	1,136.68021
1	644.94247	1	768.49466	1	892.04686	1	1,015.59906	1	1,139.15125
2	647.41351	2	770.96571	2	894.51790	2	1,018.07010	2	1,141.62230
3	649.88455	3	773.43675	3	896.98895	3	1,020.54114	3	1,144.09334
4	652.35560	4	775.90779	4	899.45999	4	1,023.01219	4	1,146.56438
5	654.82664	5	778.37884	5	901.93103	5	1,025.48323	5	1,149.03543
6	657.29769	6	780.84988	6	904.40208	6	1,027.95427	6	1,151.50647
7	659.76873	7	783.32093	7	906.87312	7	1,030.42532	7	1,153.97752
8	662.23977	8	785.79197	8	909.34417	8	1,032.89636	8	1,156.44856
9	664.71082	9	788.26301	9	911.81521	9	1,035.36741	9	1,158.91960
270	667.18186	320	790.73406	370	914.28625	420	1,037.83845	470	1,161.39065
1	669.65291	1	793.20510	1	916.75730	1	1,040.30949	1	1,163.86169
2	672.12395	2	795.67615	2	919.22834	2	1,042.78054	2	1,166.33273
3	674.59499	3	798.14719	3	921.69939	3	1,045.25158	3	1,168.80378
4	677.06604	4	800.61823	4	924.17043	4	1,047.72263	4	1,171.27482
5	679.53708	5	803.08928	5	926.64147	5	1,050.19367	5	1,173.74587
6	682.00812	6	805.56032	6	929.11252	6	1,052.66471	6	1,176.21691
7	684.47917	7	808.03137	7	931.58356	7	1,055.13576	7	1,178.68795
8	686.95021	8	810.50241	8	934.05461	8	1,057.60680	8	1,181.15900
9	689.42126	9	812.97345	9	936.52565	9	1,060.07785	9	1,183.63004
280	691.89230	330	815.44450	380	928.99669	430	1,062.54889	480	1,186.10109
1	694.36334	1	817.91554	1	941.46774	1	1,065.01993	1	1,188.57213
2	696.83439	2	820.38658	2	943.93878	2	1,067.49098	2	1,191.04317
3	699.30543	3	822.85763	3	946.40983	3	1,069.96202	3	1,193.51422
4	701.77648	4	825.32867	4	948.88087	4	1,072.43307	4	1,195.98526
5	704.24752	5	827.79972	5	951.35191	5	1,074.90411	5	1,198.45631
6	706.71856	6	830.27076	6	953.82296	6	1,077.37515	6	1,200.92735
7	709.18961	7	832.74180	7	956.29400	7	1,079.84620	7	1,203.39839
8	711.66065	8	835.21285	8	958.76504	8	1,082.31724	8	1,205.86944
9	714.13170	9	837.68389	9	961.23609	9	1,084.78829	9	1,208.34048
290	716.60274	340	840.15494	390	963.70713	440	1,087.25933	490	1,210.81153
1	719.07378	1	842.62598	1	966.17818	1	1,089.73037	1	1,213.28257
2	721.54483	2	845.09702	2	968.64922	2	1,092.20142	2	1,215.75361
3	724.01587	3	847.56807	3	971.12026	3	1,094.67246	3	1,218.22466
4	726.48692	4	850.03911	4	973.59131	4	1,097.14350	4	1,220.69570
5	728.95796	5	852.51016	5	976.06235	5	1,099.61455	5	1,223.16675
6	731.42900	6	854.98120	6	978.53340	6	1,102.08559	6	1,225.63779
7	733.90005	7	857.45224	7	981.00444	7	1,104.55664	7	1,228.10883
8	736.37109	8	859.92329	8	983.47548	8	1,107.02768	8	1,230.57988
9	738.84214	9	862.39433	9	985.94653	9	1,109.49872	9	1,233.05092

AREA—HECTARES TO ACRES

[Reduction factor: 1 hectare = 2.471043930 acres]

Hectares	Acres	Hectares	Acres	Hectares	Acres	Hectares	Acres	Hectares	Acres
500	1,235. 52197	550	1,359. 07416	600	1,482. 62636	650	1,606. 17855	700	1,729. 73075
1	1,237. 99301	1	1,361. 54521	1	1,485. 09740	1	1,608. 64960	1	1,732. 20180
2	1,240. 46405	2	1,364. 01625	2	1,487. 56845	2	1,611. 12064	2	1,734. 67284
3	1,242. 93510	3	1,366. 48729	3	1,490. 03949	3	1,613. 59169	3	1,737. 14388
4	1,245. 40614	4	1,368. 95834	4	1,492. 51053	4	1,616. 06273	4	1,739. 61493
5	1,247. 87718	5	1,371. 42938	5	1,494. 98158	5	1,618. 53377	5	1,742. 08597
6	1,250. 34823	6	1,373. 90043	6	1,497. 45262	6	1,621. 00482	6	1,744. 55701
7	1,252. 81927	7	1,376. 37147	7	1,499. 92367	7	1,623. 47586	7	1,747. 02806
8	1,255. 29032	8	1,378. 84251	8	1,502. 39471	8	1,625. 94691	8	1,749. 49910
9	1,257. 76136	9	1,381. 31356	9	1,504. 86575	9	1,628. 41795	9	1,751. 97015
510	1,260. 23240	560	1,383. 78460	610	1,507. 33680	660	1,630. 88899	710	1,754. 44119
1	1,262. 70345	1	1,386. 25564	1	1,509. 80784	1	1,633. 36004	1	1,756. 91223
2	1,265. 17449	2	1,388. 72669	2	1,512. 27889	2	1,635. 83108	2	1,759. 38328
3	1,267. 64554	3	1,391. 19773	3	1,514. 74993	3	1,638. 30213	3	1,761. 85432
4	1,270. 11658	4	1,393. 66878	4	1,517. 22097	4	1,640. 77317	4	1,764. 32537
5	1,272. 58762	5	1,396. 13982	5	1,519. 69202	5	1,643. 24421	5	1,766. 79641
6	1,275. 05867	6	1,398. 61086	6	1,522. 16306	6	1,645. 71526	6	1,769. 26745
7	1,277. 52971	7	1,401. 08191	7	1,524. 63411	7	1,648. 18630	7	1,771. 73850
8	1,280. 00076	8	1,403. 55295	8	1,527. 10515	8	1,650. 65735	8	1,774. 20954
9	1,282. 47180	9	1,406. 02400	9	1,529. 57619	9	1,653. 12839	9	1,776. 68059
520	1,284. 94284	570	1,408. 49504	620	1,532. 04724	670	1,655. 59943	720	1,779. 15163
1	1,287. 41389	1	1,410. 96608	1	1,534. 51828	1	1,658. 07048	1	1,781. 62267
2	1,289. 88493	2	1,413. 43713	2	1,536. 98932	2	1,660. 54152	2	1,784. 09372
3	1,292. 35598	3	1,415. 90817	3	1,539. 46037	3	1,663. 01257	3	1,786. 56476
4	1,294. 82702	4	1,418. 37922	4	1,541. 93141	4	1,665. 48361	4	1,789. 03581
5	1,297. 29806	5	1,420. 85026	5	1,544. 40246	5	1,667. 95465	5	1,791. 50685
6	1,299. 76911	6	1,423. 32130	6	1,546. 87350	6	1,670. 42570	6	1,793. 97789
7	1,302. 24015	7	1,425. 79235	7	1,549. 34454	7	1,672. 89674	7	1,796. 44894
8	1,304. 71120	8	1,428. 26339	8	1,551. 81559	8	1,675. 36778	8	1,798. 91998
9	1,307. 18224	9	1,430. 73444	9	1,554. 28663	9	1,677. 83883	9	1,801. 39103
530	1,309. 65328	580	1,433. 20548	630	1,556. 75768	680	1,680. 30987	730	1,803. 86207
1	1,312. 12433	1	1,435. 67652	1	1,559. 22872	1	1,682. 78092	1	1,806. 33311
2	1,314. 59537	2	1,438. 14757	2	1,561. 69976	2	1,685. 25196	2	1,808. 80416
3	1,317. 06641	3	1,440. 61861	3	1,564. 17081	3	1,687. 72300	3	1,811. 27520
4	1,319. 53746	4	1,443. 08966	4	1,566. 64185	4	1,690. 19405	4	1,813. 74624
5	1,322. 00850	5	1,445. 56070	5	1,569. 11290	5	1,692. 66509	5	1,816. 21729
6	1,324. 47955	6	1,448. 03174	6	1,571. 58394	6	1,695. 13614	6	1,818. 68833
7	1,326. 95059	7	1,450. 50279	7	1,574. 05498	7	1,697. 60718	7	1,821. 15938
8	1,329. 42163	8	1,452. 97383	8	1,576. 52603	8	1,700. 07822	8	1,823. 63042
9	1,331. 89268	9	1,455. 44487	9	1,578. 99707	9	1,702. 54927	9	1,826. 10146
540	1,334. 36372	590	1,457. 91592	640	1,581. 46812	690	1,705. 02031	740	1,828. 57251
1	1,336. 83477	1	1,460. 38696	1	1,583. 93916	1	1,707. 49136	1	1,831. 04355
2	1,339. 30581	2	1,462. 85801	2	1,586. 41020	2	1,709. 96240	2	1,833. 51460
3	1,341. 77685	3	1,465. 32905	3	1,588. 88125	3	1,712. 43344	3	1,835. 98564
4	1,344. 24790	4	1,467. 80009	4	1,591. 35229	4	1,714. 90449	4	1,838. 45668
5	1,346. 71894	5	1,470. 27114	5	1,593. 82334	5	1,717. 37553	5	1,840. 92773
6	1,349. 18999	6	1,472. 74218	6	1,596. 29438	6	1,719. 84658	6	1,843. 39877
7	1,351. 66103	7	1,475. 21323	7	1,598. 76542	7	1,722. 31762	7	1,845. 86982
8	1,354. 13207	8	1,477. 68427	8	1,601. 23647	8	1,724. 78866	8	1,848. 34086
9	1,356. 60312	9	1,480. 15531	9	1,603. 70751	9	1,727. 25971	9	1,850. 81190

AREA—HECTARES TO ACRES

[Reduction factor: 1 hectare = 2.471043930 acres]

Hectares	Acres	Hectares	Acres	Hectares	Acres	Hectares	Acres	Hectares	Acres
750	1,853.28295	800	1,976.83514	850	2,100.38734	900	2,223.93954	950	2,347.49173
1	1,855.75399	1	1,979.30619	1	2,102.85838	1	2,226.41058	1	2,349.96278
2	1,858.22504	2	1,981.77723	2	2,105.32943	2	2,228.88163	2	2,352.43382
3	1,860.69608	3	1,984.24828	3	2,107.80047	3	2,231.35267	3	2,354.90487
4	1,863.16712	4	1,986.71932	4	2,110.27152	4	2,233.82371	4	2,357.37591
5	1,865.63817	5	1,989.19036	5	2,112.74256	5	2,236.29476	5	2,359.84695
6	1,868.10921	6	1,991.66141	6	2,115.21360	6	2,238.76580	6	2,362.31800
7	1,870.58026	7	1,994.13245	7	2,117.68465	7	2,241.23684	7	2,364.78904
8	1,873.05130	8	1,996.60350	8	2,120.15569	8	2,243.70789	8	2,367.26009
9	1,875.52234	9	1,999.07454	9	2,122.62674	9	2,246.17893	9	2,369.73113
760	1,877.99339	810	2,001.54558	860	2,125.09778	910	2,248.64998	960	2,372.20217
1	1,880.46443	1	2,004.01663	1	2,127.56882	1	2,251.12102	1	2,374.67322
2	1,882.93547	2	2,006.48767	2	2,130.03987	2	2,253.59206	2	2,377.14426
3	1,885.40652	3	2,008.95872	3	2,132.51091	3	2,256.06311	3	2,379.61530
4	1,887.87756	4	2,011.42976	4	2,134.98196	4	2,258.53415	4	2,382.08635
5	1,890.34861	5	2,013.90080	5	2,137.45300	5	2,261.00520	5	2,384.55739
6	1,892.81965	6	2,016.37185	6	2,139.92404	6	2,263.47624	6	2,387.02844
7	1,895.29069	7	2,018.84289	7	2,142.39509	7	2,265.94728	7	2,389.49948
8	1,897.76174	8	2,021.31394	8	2,144.86613	8	2,268.41833	8	2,391.97052
9	1,900.23278	9	2,023.78498	9	2,147.33718	9	2,270.88937	9	2,394.44157
770	1,902.70383	820	2,026.25602	870	2,149.80822	920	2,273.36042	970	2,396.91261
1	1,905.17487	1	2,028.72707	1	2,152.27926	1	2,275.83146	1	2,399.38366
2	1,907.64591	2	2,031.19811	2	2,154.75031	2	2,278.30250	2	2,401.85470
3	1,910.11696	3	2,033.66915	3	2,157.22135	3	2,280.77355	3	2,404.32574
4	1,912.58800	4	2,036.14020	4	2,159.69240	4	2,283.24459	4	2,406.79679
5	1,915.05905	5	2,038.61124	5	2,162.16344	5	2,285.71564	5	2,409.26783
6	1,917.53009	6	2,041.08229	6	2,164.63448	6	2,288.18668	6	2,411.73888
7	1,920.00113	7	2,043.55333	7	2,167.10553	7	2,290.65772	7	2,414.20992
8	1,922.47218	8	2,046.02437	8	2,169.57657	8	2,293.12877	8	2,416.68096
9	1,924.94322	9	2,048.49542	9	2,172.04761	9	2,295.59981	9	2,419.15201
780	1,927.41427	830	2,050.96646	880	2,174.51866	930	2,298.07086	980	2,421.62305
1	1,929.88531	1	2,053.43751	1	2,176.98970	1	2,300.54190	1	2,424.09410
2	1,932.35635	2	2,055.90855	2	2,179.46075	2	2,303.01294	2	2,426.56514
3	1,934.82740	3	2,058.37959	3	2,181.93179	3	2,305.48399	3	2,429.03618
4	1,937.29844	4	2,060.85064	4	2,184.40283	4	2,307.95503	4	2,431.50723
5	1,939.76949	5	2,063.32168	5	2,186.87388	5	2,310.42607	5	2,433.97827
6	1,942.24053	6	2,065.79273	6	2,189.34492	6	2,312.89712	6	2,436.44932
7	1,944.71157	7	2,068.26377	7	2,191.81597	7	2,315.36816	7	2,438.92036
8	1,947.18262	8	2,070.73481	8	2,194.28701	8	2,317.83921	8	2,441.39140
9	1,949.65366	9	2,073.20586	9	2,196.75805	9	2,320.31025	9	2,443.86245
790	1,952.12471	840	2,075.67690	890	2,199.22910	940	2,322.78129	990	2,446.33349
1	1,954.59575	1	2,078.14795	1	2,201.70014	1	2,325.25234	1	2,448.80454
2	1,957.06679	2	2,080.61899	2	2,204.17119	2	2,327.72338	2	2,451.27558
3	1,959.53784	3	2,083.09003	3	2,206.64223	3	2,330.19443	3	2,453.74662
4	1,962.00888	4	2,085.56108	4	2,209.11327	4	2,332.66547	4	2,456.21767
5	1,964.47992	5	2,088.03212	5	2,211.58432	5	2,335.13651	5	2,458.68871
6	1,966.95097	6	2,090.50317	6	2,214.05536	6	2,337.60756	6	2,461.15975
7	1,969.42201	7	2,092.97421	7	2,216.52641	7	2,340.07860	7	2,463.63080
8	1,971.89305	8	2,095.44525	8	2,218.99745	8	2,342.54965	8	2,466.10184
9	1,974.36410	9	2,097.91630	9	2,221.46849	9	2,345.02069	9	2,468.57289

AREA—ACRES TO HECTARES

[Reduction factor: 1 acre = 0.4046872610 hectare]

Acres	Hectares	Acres	Hectares	Acres	Hectares	Acres	Hectares	Acres	Hectares
0		50	20.23436	100	40.46873	150	60.70309	200	80.93745
1	0.40469	1	20.63905	1	40.87341	1	61.10778	1	81.34214
2	0.80937	2	21.04374	2	41.27810	2	61.51246	2	81.74683
3	1.21406	3	21.44842	3	41.68279	3	61.91715	3	82.15151
4	1.61875	4	21.85311	4	42.08748	4	62.32184	4	82.55620
5	2.02344	5	22.25780	5	42.49216	5	62.72653	5	82.96089
6	2.42812	6	22.66249	6	42.89685	6	63.13121	6	83.36558
7	2.83281	7	23.06717	7	43.30154	7	63.53590	7	83.77026
8	3.23750	8	23.47186	8	43.70622	8	63.94059	8	84.17495
9	3.64219	9	23.87655	9	44.11091	9	64.34527	9	84.57964
10	4.04688	60	24.28124	110	44.51560	160	64.74996	210	84.98432
1	4.45156	1	24.68592	1	44.92029	1	65.15465	1	85.38901
2	4.85625	2	25.09061	2	45.32497	2	65.55934	2	85.79370
3	5.26093	3	25.49530	3	45.72966	3	65.96402	3	86.19839
4	5.66562	4	25.89998	4	46.13435	4	66.36871	4	86.60307
5	6.07031	5	26.30467	5	46.53904	5	66.77340	5	87.00776
6	6.47500	6	26.70936	6	46.94372	6	67.17809	6	87.41245
7	6.87968	7	27.11405	7	47.34841	7	67.58277	7	87.81714
8	7.28437	8	27.51873	8	47.75310	8	67.98746	8	88.22182
9	7.68906	9	27.92342	9	48.15778	9	68.39215	9	88.62651
20	8.09375	70	28.32811	120	48.56247	170	68.79683	220	89.03120
1	8.49843	1	28.73280	1	48.96716	1	69.20152	1	89.43588
2	8.90312	2	29.13748	2	49.37185	2	69.60621	2	89.84057
3	9.30781	3	29.54217	3	49.77653	3	70.01090	3	90.24526
4	9.71249	4	29.94686	4	50.18122	4	70.41558	4	90.64995
5	10.11718	5	30.35154	5	50.58591	5	70.82027	5	91.05463
6	10.52187	6	30.75623	6	50.99059	6	71.22496	6	91.45932
7	10.92656	7	31.16092	7	51.39528	7	71.62965	7	91.86401
8	11.33124	8	31.56561	8	51.79997	8	72.03433	8	92.26870
9	11.73593	9	31.97029	9	52.20466	9	72.43902	9	92.67338
30	12.14062	80	32.37498	130	52.60934	180	72.84371	230	93.07807
1	12.54531	1	32.77967	1	53.01403	1	73.24839	1	93.48276
2	12.94999	2	33.18436	2	53.41872	2	73.65308	2	93.88744
3	13.35468	3	33.58904	3	53.82341	3	74.05777	3	94.29213
4	13.75937	4	33.99373	4	54.22809	4	74.46246	4	94.69682
5	14.16405	5	34.39842	5	54.63278	5	74.86714	5	95.10151
6	14.56874	6	34.80310	6	55.03747	6	75.27183	6	95.50619
7	14.97343	7	35.20779	7	55.44215	7	75.67652	7	95.91088
8	15.37812	8	35.61248	8	55.84684	8	76.08121	8	96.31557
9	15.78280	9	36.01717	9	56.25153	9	76.48589	9	96.72026
40	16.18749	90	36.42185	140	56.65622	190	76.89058	240	97.12494
1	16.59218	1	36.82654	1	57.06090	1	77.29527	1	97.52963
2	16.99686	2	37.23123	2	57.46559	2	77.69995	2	97.93432
3	17.40155	3	37.63592	3	57.87028	3	78.10464	3	98.33900
4	17.80624	4	38.04060	4	58.27497	4	78.50933	4	98.74369
5	18.21093	5	38.44529	5	58.67965	5	78.91402	5	99.14838
6	18.61561	6	38.84998	6	59.08434	6	79.31870	6	99.55307
7	19.02030	7	39.25466	7	59.48903	7	79.72339	7	99.95775
8	19.42499	8	39.65935	8	59.89371	8	80.12808	8	100.36244
9	19.82968	9	40.06404	9	60.29840	9	80.53276	9	100.76713

AREA—ACRES TO HECTARES

[Reduction factor: 1 acre = 0.4046872610 hectare]

Acres	Hectares	Acres	Hectares	Acres	Hectares	Acres	Hectares	Acres	Hectares
250	101.17182	300	121.40618	350	141.64054	400	161.87490	450	182.10927
1	101.57650	1	121.81087	1	142.04523	1	162.27959	1	182.51395
2	101.98119	2	122.21555	2	142.44992	2	162.68428	2	182.91864
3	102.38588	3	122.62024	3	142.85460	3	163.08897	3	183.32333
4	102.79056	4	123.02493	4	143.25929	4	163.49365	4	183.72802
5	103.19525	5	123.42961	5	143.66398	5	163.89834	5	184.13270
6	103.59994	6	123.83430	6	144.06866	6	164.30303	6	184.53739
7	104.00463	7	124.23899	7	144.47335	7	164.70772	7	184.94208
8	104.40931	8	124.64368	8	104.87804	8	165.11240	8	185.34677
9	104.81400	9	125.04836	9	145.28273	9	165.51709	9	185.75145
260	105.21869	310	125.45305	360	145.68741	410	165.92178	460	186.15614
1	105.62338	1	125.85774	1	146.09210	1	166.32646	1	186.56083
2	106.02806	2	126.26243	2	146.49679	2	166.73115	2	186.96551
3	106.43275	3	126.66711	3	146.90148	3	167.13584	3	187.37020
4	106.83744	4	127.07180	4	147.30616	4	167.54053	4	187.77489
5	107.24212	5	127.47649	5	147.71085	5	167.94521	5	188.17958
6	107.64681	6	127.88117	6	148.11554	6	168.34990	6	188.58426
7	108.05150	7	128.28586	7	148.52022	7	168.75459	7	188.98895
8	108.45619	8	128.69055	8	148.92491	8	169.15928	8	189.39364
9	108.86087	9	129.09524	9	149.32960	9	169.56396	9	189.79833
270	109.26556	320	129.49992	370	149.73429	420	169.96865	470	190.20301
1	109.67025	1	129.90461	1	150.13897	1	170.37334	1	190.60770
2	110.07493	2	130.30930	2	150.54366	2	170.77802	2	191.01239
3	110.47962	3	130.71399	3	150.94835	3	171.18271	3	191.41707
4	110.88431	4	131.11867	4	151.35304	4	171.58740	4	191.82176
5	111.28900	5	131.52336	5	151.75772	5	171.99209	5	192.22645
6	111.69368	6	131.92805	6	152.16241	6	172.39677	6	192.63114
7	112.09837	7	132.33273	7	152.56710	7	172.80146	7	193.03582
8	112.50306	8	132.73742	8	152.97178	8	173.20615	8	193.44051
9	112.90775	9	133.14211	9	153.37647	9	173.61083	9	193.84520
280	113.31243	330	133.54680	380	153.78116	430	174.01552	480	194.24989
1	113.71712	1	133.95148	1	154.18585	1	174.42021	1	194.65457
2	114.12181	2	134.35617	2	154.59053	2	174.82490	2	195.05926
3	114.52649	3	134.76086	3	154.99522	3	175.22958	3	195.46395
4	114.93118	4	135.16555	4	155.39991	4	175.63427	4	195.86863
5	115.33587	5	135.57023	5	155.80460	5	176.03896	5	196.27332
6	115.74056	6	135.97492	6	156.20928	6	176.44365	6	196.67801
7	116.14524	7	136.37961	7	156.61397	7	176.84833	7	197.08270
8	116.54993	8	136.78429	8	157.01866	8	177.25302	8	197.48738
9	116.95462	9	137.18898	9	157.42334	9	177.65771	9	197.89207
290	117.35931	340	137.59367	390	157.82803	440	178.06239	490	198.29676
1	117.76399	1	137.99836	1	158.23272	1	178.46708	1	198.70145
2	118.16868	2	138.40304	2	158.63741	2	178.87177	2	199.10613
3	118.57337	3	138.80773	3	159.04209	3	179.27646	3	199.51082
4	118.97805	4	139.21242	4	159.44678	4	179.68114	4	199.91551
5	119.38274	5	139.61711	5	159.85147	5	180.08583	5	200.32019
6	119.78743	6	140.02179	6	160.25616	6	180.49052	6	200.72488
7	120.19212	7	140.42648	7	160.66084	7	180.89521	7	201.12957
8	120.59680	8	140.83117	8	161.06553	8	181.29989	8	201.53426
9	121.00149	9	141.23585	9	161.47022	9	181.70458	9	201.93894

AREA—ACRES TO HECTARES

[Reduction factor: 1 acre = 0.4046872610 hectare]

Acres	Hectares	Acres	Hectares	Acres	Hectares	Acres	Hectares	Acres	Hectares
500	202.34363	550	222.57799	600	242.81236	650	263.04672	700	283.28108
1	202.74832	1	222.98268	1	243.21704	1	263.45141	1	283.68577
2	203.15301	2	223.38737	2	243.62173	2	263.85609	2	284.09046
3	203.55769	3	223.79206	3	244.02642	3	264.26078	3	284.49514
4	203.96238	4	224.19674	4	244.43111	4	264.66547	4	284.89983
5	204.36707	5	224.60143	5	244.83579	5	265.07016	5	285.30452
6	204.77175	6	225.00612	6	245.24048	6	265.47484	6	285.70921
7	205.17644	7	225.41080	7	245.64517	7	265.87953	7	286.11389
8	205.58113	8	225.81549	8	246.04985	8	266.28422	8	286.51858
9	205.98582	9	226.22018	9	246.45454	9	266.68890	9	286.92327
510	206.39050	560	226.62487	610	246.85923	660	267.09359	710	287.32796
1	206.79519	1	227.02955	1	247.26392	1	267.49828	1	287.73264
2	207.19988	2	227.43424	2	247.66860	2	267.90297	2	288.13733
3	207.60456	3	227.83893	3	248.07329	3	268.30765	3	288.54202
4	208.00925	4	228.24362	4	248.47798	4	268.71234	4	288.94670
5	208.41394	5	228.64830	5	248.88267	5	269.11703	5	289.35139
6	208.81863	6	229.05299	6	249.28735	6	269.52172	6	289.75608
7	209.22331	7	229.45768	7	249.69204	7	269.92640	7	290.16077
8	209.62800	8	229.86236	8	250.09673	8	270.33109	8	290.56545
9	210.03269	9	230.26705	9	250.50141	9	270.73578	9	290.97014
520	210.43738	570	230.67174	620	250.90610	670	271.14046	720	291.37483
1	210.84206	1	231.07643	1	251.31079	1	271.54515	1	291.77952
2	211.24675	2	231.48111	2	251.71548	2	271.94984	2	292.18420
3	211.65144	3	231.88580	3	252.12016	3	272.35453	3	292.58889
4	212.05612	4	232.29049	4	252.52485	4	272.75921	4	292.99358
5	212.46081	5	232.69518	5	252.92954	5	273.16390	5	293.39826
6	212.86550	6	233.09986	6	253.33423	6	273.56859	6	293.80295
7	213.27019	7	233.50455	7	253.73891	7	273.97328	7	294.20764
8	213.67487	8	233.90924	8	254.14360	8	274.37796	8	294.61233
9	214.07956	9	234.31392	9	254.54829	9	274.78265	9	295.01701
530	214.48425	580	234.71861	630	254.95297	680	275.18734	730	295.42170
1	214.88894	1	235.12330	1	255.35766	1	275.59202	1	295.82639
2	215.29362	2	235.52799	2	255.76235	2	275.99671	2	296.23108
3	215.69831	3	235.93267	3	256.16704	3	276.40140	3	296.63576
4	216.10300	4	236.33736	4	256.57172	4	276.80609	4	297.04045
5	216.50768	5	236.74205	5	256.97641	5	277.21077	5	297.44514
6	216.91237	6	237.14673	6	257.38110	6	277.61546	6	297.84982
7	217.31706	7	237.55142	7	257.78579	7	278.02015	7	298.25451
8	217.72175	8	237.95611	8	258.19047	8	278.42484	8	298.65920
9	218.12643	9	238.36080	9	258.59516	9	278.82952	9	299.06389
540	218.53112	590	238.76548	640	258.99985	690	279.23421	740	299.46857
1	218.93581	1	239.17017	1	259.40453	1	279.63890	1	299.87326
2	219.34050	2	239.57486	2	259.80922	2	280.04358	2	300.27795
3	219.74518	3	239.97955	3	260.21391	3	280.44827	3	300.68263
4	220.14987	4	240.38423	4	260.61860	4	280.85296	4	301.08732
5	220.55456	5	240.78892	5	261.02328	5	281.25765	5	301.49201
6	220.95924	6	241.19361	6	261.42797	6	281.66233	6	301.89670
7	221.36393	7	241.59829	7	261.83266	7	282.06702	7	302.30138
8	221.76862	8	242.00298	8	262.23735	8	282.47171	8	302.70607
9	222.17331	9	242.40767	9	262.64203	9	282.87640	9	303.11076

AREA—ACRES TO HECTARES

[Reduction factor: 1 acre = 0.4046872610 hectare]

Acres	Hectares	Acres	Hectares	Acres	Hectares	Acres	Hectares	Acres	Hectares
750	303.51545	800	323.74981	850	343.98417	900	364.21853	950	384.45290
1	303.92013	1	324.15450	1	344.38886	1	364.62322	1	384.85759
2	304.32482	2	324.55918	2	344.79355	2	365.02791	2	385.26227
3	304.72951	3	324.96387	3	345.19823	3	365.43260	3	385.66696
4	305.13419	4	325.36856	4	345.60292	4	365.83728	4	386.07165
5	305.53888	5	325.77325	5	346.00761	5	366.24197	5	386.47633
6	305.94357	6	326.17793	6	346.41230	6	366.64666	6	386.88102
7	306.34826	7	326.58262	7	346.81698	7	367.05135	7	387.28571
8	306.75294	8	326.98731	8	347.22167	8	367.45603	8	387.69040
9	307.15763	9	327.39199	9	347.62636	9	367.86072	9	388.09508
760	307.56232	810	327.79668	860	348.03104	910	368.26541	960	388.49977
1	307.96701	1	328.20137	1	348.43573	1	368.67009	1	388.90446
2	308.37169	2	328.60606	2	348.84042	2	369.07478	2	389.30915
3	308.77638	3	329.01074	3	349.24511	3	369.47947	3	389.71383
4	309.18107	4	329.41543	4	349.64979	4	369.88416	4	390.11852
5	309.58575	5	329.82012	5	350.05448	5	370.28884	5	390.52321
6	309.99044	6	330.22480	6	350.45917	6	370.69353	6	390.92789
7	310.39513	7	330.62949	7	350.86386	7	371.09822	7	391.33258
8	310.79982	8	331.03418	8	351.26854	8	371.50291	8	391.73727
9	311.20450	9	331.43887	9	351.67323	9	371.90759	9	392.14196
770	311.60919	820	331.84355	870	352.07792	920	372.31228	970	392.54664
1	312.01388	1	332.24824	1	352.48260	1	372.71697	1	392.95133
2	312.41857	2	332.65293	2	352.88729	2	373.12165	2	393.35602
3	312.82325	3	333.05762	3	353.29198	3	373.52634	3	393.76070
4	313.22794	4	333.46230	4	353.69667	4	373.93103	4	394.16539
5	313.63263	5	333.86699	5	354.10135	5	374.33572	5	394.57008
6	314.03731	6	334.27168	6	354.50604	6	374.74040	6	394.97477
7	314.44200	7	334.67636	7	354.91073	7	375.14509	7	395.37945
8	314.84669	8	335.08105	8	355.31542	8	375.54978	8	395.78414
9	315.25138	9	335.48574	9	355.72010	9	375.95447	9	396.18883
780	315.65606	830	335.89043	880	356.12479	930	376.35915	980	396.59352
1	316.06075	1	336.29511	1	356.52948	1	376.76384	1	396.99820
2	316.46544	2	336.69980	2	356.93416	2	377.16853	2	397.40289
3	316.87013	3	337.10449	3	357.33885	3	377.57321	3	397.80758
4	317.27481	4	337.50918	4	357.74354	4	377.97790	4	398.21226
5	317.67950	5	337.91386	5	358.14823	5	378.38259	5	398.61695
6	318.08419	6	338.31855	6	358.55291	6	378.78728	6	399.02164
7	318.48887	7	338.72324	7	358.95760	7	379.19196	7	399.42633
8	318.89356	8	339.12792	8	359.36229	8	379.59665	8	399.83101
9	319.29825	9	339.53261	9	359.76698	9	380.00134	9	400.23570
790	319.70294	840	339.93730	890	360.17166	940	380.40603	990	400.64039
1	320.10762	1	340.34199	1	360.57635	1	380.81071	1	401.04508
2	320.51231	2	340.74667	2	360.98104	2	381.21540	2	401.44976
3	320.91700	3	341.15136	3	361.38572	3	381.62009	3	401.85445
4	321.32169	4	341.55605	4	361.79041	4	382.02477	4	402.25914
5	321.72637	5	341.96074	5	362.19510	5	382.42946	5	402.66382
6	322.13106	6	342.36542	6	362.59979	6	382.83415	6	403.06851
7	322.53575	7	342.77011	7	363.00447	7	383.23884	7	403.47320
8	322.94043	8	343.17480	8	363.40916	8	383.64352	8	403.87789
9	323.34512	9	343.57948	9	363.81385	9	384.04821	9	404.28257

VOLUME—CUBIC METRES TO CUBIC YARDS

[Reduction factor: 1 cubic metre = 1.307942772 cubic yards]

Cubic metres	Cubic yards	Cubic metres	Cubic yards	Cubic metres	Cubic yards	Cubic metres	Cubic yards	Cubic metres	Cubic yards
0		50	65.39714	100	130.79428	150	196.19142	200	261.58855
1	1.30794	1	66.70508	1	132.10222	1	197.49936	1	262.89650
2	2.61589	2	68.01302	2	133.41016	2	198.80730	2	264.20444
3	3.92383	3	69.32097	3	134.71811	3	200.11524	3	265.51238
4	5.23177	4	70.62891	4	136.02605	4	201.42319	4	266.82033
5	6.53971	5	71.93685	5	137.33399	5	202.73113	5	268.12827
6	7.84766	6	73.24480	6	138.64193	6	204.03907	6	269.43621
7	9.15560	7	74.55274	7	139.94988	7	205.34702	7	270.74415
8	10.46354	8	75.86068	8	141.25782	8	206.65496	8	272.05210
9	11.77148	9	77.16862	9	142.56576	9	207.96290	9	273.36004
10	13.07943	60	78.47657	110	143.87370	160	209.27084	210	274.66798
1	14.38737	1	79.78451	1	145.18165	1	210.57879	1	275.97592
2	15.69531	2	81.09245	2	146.48959	2	211.88673	2	277.28387
3	17.00326	3	82.40039	3	147.79753	3	213.19467	3	278.59181
4	18.31120	4	83.70834	4	149.10548	4	214.50261	4	279.89975
5	19.61914	5	85.01628	5	150.41342	5	215.81056	5	281.20770
6	20.92708	6	86.32422	6	151.72136	6	217.11850	6	282.51564
7	22.23503	7	87.63217	7	153.02930	7	218.42644	7	283.82358
8	23.54297	8	88.94011	8	154.33725	8	219.73439	8	285.13152
9	24.85091	9	90.24805	9	155.64519	9	221.04233	9	286.43947
20	26.15886	70	91.55599	120	156.95313	170	222.35027	220	287.74741
1	27.46680	1	92.86394	1	158.26108	1	223.65821	1	289.05535
2	28.77474	2	94.17188	2	159.56902	2	224.96616	2	290.36330
3	30.08268	3	95.47982	3	160.87696	3	226.27410	3	291.67124
4	31.39063	4	96.78777	4	162.18490	4	227.58204	4	292.97918
5	32.69857	5	98.09571	5	163.49285	5	228.88999	5	294.28712
6	34.00651	6	99.40365	6	164.80079	6	230.19793	6	295.59507
7	35.31445	7	100.71159	7	166.10873	7	231.50587	7	296.90301
8	36.62240	8	102.01954	8	167.41667	8	232.81381	8	298.21095
9	37.93034	9	103.32748	9	168.72462	9	234.12176	9	299.51889
30	39.23828	80	104.63542	130	170.03256	180	235.42970	230	300.82684
1	40.54623	1	105.94336	1	171.34050	1	236.73764	1	302.13478
2	41.85417	2	107.25131	2	172.64845	2	238.04558	2	303.44272
3	43.16211	3	108.55925	3	173.95639	3	239.35353	3	304.75067
4	44.47005	4	109.86719	4	175.26433	4	240.66147	4	306.05861
5	45.77800	5	111.17514	5	176.57227	5	241.96941	5	307.36655
6	47.08594	6	112.48308	6	177.88022	6	243.27736	6	308.67449
7	48.39388	7	113.79102	7	179.18816	7	244.58530	7	309.98244
8	49.70183	8	115.09896	8	180.49610	8	245.89324	8	311.29038
9	51.00977	9	116.40691	9	181.80405	9	247.20118	9	312.59832
40	52.31771	90	117.71485	140	183.11199	190	248.50913	240	313.90627
1	53.62565	1	119.02279	1	184.41993	1	249.81707	1	315.21421
2	54.93360	2	120.33074	2	185.72787	2	251.12501	2	316.52215
3	56.24154	3	121.63868	3	187.03582	3	252.43295	3	317.83009
4	57.54948	4	122.94662	4	188.34376	4	253.74090	4	319.13804
5	58.85742	5	124.25456	5	189.65170	5	255.04884	5	320.44598
6	60.16537	6	125.56251	6	190.95964	6	256.35678	6	321.75392
7	61.47331	7	126.87045	7	192.26759	7	257.66473	7	323.06186
8	62.78125	8	128.17839	8	193.57553	8	258.97267	8	324.36981
9	64.08920	9	129.48633	9	194.88347	9	260.28061	9	325.67775

VOLUME—CUBIC METRES TO CUBIC YARDS

[Reduction factor: 1 cubic metre = 1.307942772 cubic yards]

Cubic metres	Cubic yards	Cubic metres	Cubic yards	Cubic metres	Cubic yards	Cubic metres	Cubic yards	Cubic metres	Cubic yards
250	326.98569	300	392.38283	350	457.77997	400	523.17711	450	588.57425
1	328.29364	1	393.69077	1	459.08791	1	524.48505	1	589.88219
2	329.60158	2	394.99872	2	460.39586	2	525.79299	2	591.19013
3	330.90952	3	396.30666	3	461.70380	3	527.10094	3	592.49808
4	332.21746	4	397.61460	4	463.01174	4	528.40888	4	593.80602
5	333.52541	5	398.92255	5	464.31968	5	529.71682	5	595.11396
6	334.83335	6	400.23049	6	465.62763	6	531.02477	6	596.42190
7	336.14129	7	401.53843	7	466.93557	7	532.33271	7	597.72985
8	337.44924	8	402.84637	8	468.24351	8	533.64065	8	599.03779
9	338.75718	9	404.15432	9	469.55146	9	534.94859	9	600.34573
260	340.06512	310	405.46226	360	470.85940	410	536.25654	460	601.65368
1	341.37306	1	406.77020	1	472.16734	1	537.56448	1	602.96162
2	342.68101	2	408.07814	2	473.47528	2	538.87242	2	604.26956
3	343.98895	3	409.38609	3	474.78323	3	540.18036	3	605.57750
4	345.29689	4	410.69403	4	476.09117	4	541.48831	4	606.88545
5	346.60483	5	412.00197	5	477.39911	5	542.79625	5	608.19339
6	347.91278	6	413.30992	6	478.70705	6	544.10419	6	609.50133
7	349.22072	7	414.61786	7	480.01500	7	545.41214	7	610.80927
8	350.52866	8	415.92580	8	481.32294	8	546.72008	8	612.11722
9	351.83661	9	417.23374	9	482.63088	9	548.02802	9	613.42516
270	353.14455	320	418.54169	370	483.93883	420	549.33596	470	614.73310
1	354.45249	1	419.84963	1	485.24677	1	550.64391	1	616.04105
2	355.76043	2	421.15757	2	486.55471	2	551.95185	2	617.34899
3	357.06838	3	422.46552	3	487.86265	3	553.25979	3	618.65693
4	358.37632	4	423.77346	4	489.17060	4	554.56774	4	619.96487
5	359.68426	5	425.08140	5	490.47854	5	555.87568	5	621.27282
6	360.99221	6	426.38934	6	491.78648	6	557.18362	6	622.58076
7	362.30015	7	427.69729	7	493.09443	7	558.49156	7	623.88870
8	363.60809	8	429.00523	8	494.40237	8	559.79951	8	625.19665
9	364.91603	9	430.31317	9	495.71031	9	561.10745	9	626.50459
280	366.22398	330	431.62111	380	497.01825	430	562.41539	480	627.81253
1	367.53192	1	432.92906	1	498.32620	1	563.72333	1	629.12047
2	368.83986	2	434.23700	2	499.63414	2	565.03128	2	630.42842
3	370.14780	3	435.54494	3	500.94208	3	566.33922	3	631.73636
4	371.45575	4	436.85289	4	502.25002	4	567.64716	4	633.04430
5	372.76369	5	438.16083	5	503.55797	5	568.95511	5	634.35224
6	374.07163	6	439.46877	6	504.86591	6	570.26305	6	635.66019
7	375.37958	7	440.77671	7	506.17385	7	571.57099	7	636.96813
8	376.68752	8	442.08466	8	507.48180	8	572.87893	8	638.27607
9	377.99546	9	443.39260	9	508.78974	9	574.18688	9	639.58402
290	379.30340	340	444.70054	390	510.09768	440	575.49482	490	640.89196
1	380.61135	1	446.00849	1	511.40562	1	576.80276	1	642.19990
2	381.91929	2	447.31643	2	512.71357	2	578.11071	2	643.50784
3	383.22723	3	448.62437	3	514.02151	3	579.41865	3	644.81579
4	384.53517	4	449.93231	4	515.32945	4	580.72659	4	646.12373
5	385.84312	5	451.24026	5	516.63739	5	582.03453	5	647.43167
6	387.15106	6	452.54820	6	517.94534	6	583.34248	6	648.73961
7	388.45900	7	453.85614	7	519.25328	7	584.65042	7	650.04756
8	389.76695	8	455.16408	8	520.56122	8	585.95836	8	651.35550
9	391.07489	9	456.47203	9	521.86917	9	587.26630	9	652.66344

VOLUME—CUBIC METRES TO CUBIC YARDS

[Reduction factor: 1 cubic metre = 1.307942772 cubic yards]

Cubic metres	Cubic yards	Cubic metres	Cubic yards	Cubic metres	Cubic yards	Cubic metres	Cubic yards	Cubic metres	Cubic yards
500	653.97139	550	719.36852	600	784.76566	650	850.16280	700	915.55994
1	655.27933	1	720.67647	1	786.07361	1	851.47074	1	916.86788
2	656.58727	2	721.98441	2	787.38155	2	852.77869	2	918.17583
3	657.89521	3	723.29235	3	788.68949	3	854.08663	3	919.48377
4	659.20316	4	724.60030	4	789.99743	4	855.39457	4	920.79171
5	660.51110	5	725.90824	5	791.30538	5	856.70252	5	922.09965
6	661.81904	6	727.21618	6	792.61332	6	858.01046	6	923.40760
7	663.12699	7	728.52412	7	793.92126	7	859.31840	7	924.71554
8	664.43493	8	729.83207	8	795.22921	8	860.62634	8	926.02348
9	665.74287	9	731.14001	9	796.53715	9	861.93429	9	927.33143
510	667.05081	560	732.44795	610	797.84509	660	863.24223	710	928.63937
1	668.35876	1	733.75590	1	799.15303	1	864.55017	1	929.94731
2	669.66670	2	735.06384	2	800.46098	2	865.85812	2	931.25525
3	670.97464	3	736.37178	3	801.76892	3	867.16606	3	932.56320
4	672.28258	4	737.67972	4	803.07686	4	868.47400	4	933.87114
5	673.59053	5	738.98767	5	804.38480	5	869.78194	5	935.17908
6	674.89847	6	740.29561	6	805.69275	6	871.08989	6	936.48702
7	676.20641	7	741.60355	7	807.00069	7	872.39783	7	937.79497
8	677.51436	8	742.91149	8	808.30863	8	873.70577	8	939.10291
9	678.82230	9	744.21944	9	809.61658	9	875.01371	9	940.41085
520	680.13024	570	745.52738	620	810.92452	670	876.32166	720	941.71880
1	681.43818	1	746.83532	1	812.23246	1	877.62960	1	943.02674
2	682.74613	2	748.14327	2	813.54040	2	878.93754	2	944.33468
3	684.05407	3	749.45121	3	814.84835	3	880.24549	3	945.64262
4	685.36201	4	750.75915	4	816.15629	4	881.55343	4	946.95057
5	686.66996	5	752.06709	5	817.46423	5	882.86137	5	948.25851
6	687.97790	6	753.37504	6	818.77218	6	884.16931	6	949.56645
7	689.28584	7	754.68298	7	820.08012	7	885.47726	7	950.87440
8	690.59378	8	755.99092	8	821.38806	8	886.78520	8	952.18234
9	691.90173	9	757.29886	9	822.69600	9	888.09314	9	953.49028
530	693.20967	580	758.60681	630	824.00395	680	889.40108	730	954.79822
1	694.51761	1	759.91475	1	825.31189	1	890.70903	1	956.10617
2	695.82555	2	761.22269	2	826.61983	2	892.01697	2	957.41411
3	697.13350	3	762.53064	3	827.92777	3	893.32491	3	958.72205
4	698.44144	4	763.83858	4	829.23572	4	894.63286	4	960.02999
5	699.74938	5	765.14652	5	830.54366	5	895.94080	5	961.33794
6	701.05733	6	766.45446	6	831.85160	6	897.24874	6	962.64588
7	702.36527	7	767.76241	7	833.15955	7	898.55668	7	963.95382
8	703.67321	8	769.07035	8	834.46749	8	899.86463	8	965.26177
9	704.98115	9	770.37829	9	835.77543	9	901.17257	9	966.56971
540	706.28910	590	771.68624	640	837.08337	690	902.48051	740	967.87765
1	707.59704	1	772.99418	1	838.39132	1	903.78846	1	969.18559
2	708.90498	2	774.30212	2	839.69926	2	905.09640	2	970.49354
3	710.21293	3	775.61006	3	841.00720	3	906.40434	3	971.80148
4	711.52087	4	776.91801	4	842.31515	4	907.71228	4	973.10942
5	712.82881	5	778.22595	5	843.62309	5	909.02023	5	974.41737
6	714.13675	6	779.53389	6	844.93103	6	910.32817	6	975.72531
7	715.44470	7	780.84183	7	846.23897	7	911.63611	7	977.03325
8	716.75264	8	782.14978	8	847.54692	8	912.94405	8	978.34119
9	718.06058	9	783.45772	9	848.85486	9	914.25200	9	979.64914

VOLUME—CUBIC METRES TO CUBIC YARDS

[Reduction factor: 1 cubic metre = 1.307942772 cubic yards]

Cubic metres	Cubic yards	Cubic metres	Cubic yards	Cubic metres	Cubic yards	Cubic metres	Cubic yards	Cubic metres	Cubic yards
750	980.95708	800	1,046.35422	850	1,111.75136	900	1,177.14849	950	1,242.54569
1	982.26502	1	1,047.66216	1	1,113.05930	1	1,178.45644	1	1,243.85358
2	983.57296	2	1,048.97010	2	1,114.36724	2	1,179.76438	2	1,245.16152
3	984.88091	3	1,050.27805	3	1,115.67518	3	1,181.07232	3	1,246.46946
4	986.18885	4	1,051.58599	4	1,116.98313	4	1,182.38027	4	1,247.77740
5	987.49679	5	1,052.89393	5	1,118.29107	5	1,183.68821	5	1,249.08535
6	988.80474	6	1,054.20187	6	1,119.59901	6	1,184.99615	6	1,250.39329
7	990.11268	7	1,055.50982	7	1,120.90696	7	1,186.30409	7	1,251.70123
8	991.42062	8	1,056.81776	8	1,122.21490	8	1,187.61204	8	1,253.00918
9	992.72856	9	1,058.12570	9	1,123.52284	9	1,188.91998	9	1,254.31712
760	994.03651	810	1,059.43365	860	1,124.83078	910	1,190.22792	960	1,255.62506
1	995.34445	1	1,060.74159	1	1,126.13873	1	1,191.53587	1	1,256.93300
2	996.65239	2	1,062.04953	2	1,127.44667	2	1,192.84381	2	1,258.24095
3	997.96034	3	1,063.35747	3	1,128.75461	3	1,194.15175	3	1,259.54889
4	999.26828	4	1,064.66542	4	1,130.06256	4	1,195.45969	4	1,260.85683
5	1,000.57622	5	1,065.97336	5	1,131.37050	5	1,196.76764	5	1,262.16477
6	1,001.88416	6	1,067.28130	6	1,132.67844	6	1,198.07558	6	1,263.47272
7	1,003.19211	7	1,068.58924	7	1,133.98638	7	1,199.38352	7	1,264.78066
8	1,004.50005	8	1,069.89719	8	1,135.29433	8	1,200.69146	8	1,266.08860
9	1,005.80799	9	1,071.20513	9	1,136.60227	9	1,201.99941	9	1,267.39655
770	1,007.11593	820	1,072.51307	870	1,137.91021	920	1,203.30735	970	1,268.70449
1	1,008.42388	1	1,073.82102	1	1,139.21815	1	1,204.61529	1	1,270.01243
2	1,009.73182	2	1,075.12896	2	1,140.52610	2	1,205.92324	2	1,271.32037
3	1,011.03976	3	1,076.43690	3	1,141.83404	3	1,207.23118	3	1,272.62832
4	1,012.34771	4	1,077.74484	4	1,143.14198	4	1,208.53912	4	1,273.93626
5	1,013.65565	5	1,079.05279	5	1,144.44993	5	1,209.84706	5	1,275.24420
6	1,014.96359	6	1,080.36073	6	1,145.75787	6	1,211.15501	6	1,276.55215
7	1,016.27153	7	1,081.66867	7	1,147.06581	7	1,212.46295	7	1,277.86009
8	1,017.57948	8	1,082.97662	8	1,148.37375	8	1,213.77089	8	1,279.16803
9	1,018.88742	9	1,084.28456	9	1,149.68170	9	1,215.07884	9	1,280.47597
780	1,020.19536	830	1,085.59250	880	1,150.98964	930	1,216.38678	980	1,281.78392
1	1,021.50330	1	1,086.90044	1	1,152.29758	1	1,217.69472	1	1,283.09186
2	1,022.81125	2	1,088.20839	2	1,153.60552	2	1,219.00266	2	1,284.39980
3	1,024.11919	3	1,089.51633	3	1,154.91347	3	1,220.31061	3	1,285.70774
4	1,025.42713	4	1,090.82427	4	1,156.22141	4	1,221.61855	4	1,287.01569
5	1,026.73508	5	1,092.13221	5	1,157.52935	5	1,222.92649	5	1,288.32363
6	1,028.04302	6	1,093.44016	6	1,158.83730	6	1,224.23443	6	1,289.63157
7	1,029.35096	7	1,094.74810	7	1,160.14524	7	1,225.54238	7	1,290.93952
8	1,030.65890	8	1,096.05604	8	1,161.45318	8	1,226.85032	8	1,292.24746
9	1,031.96685	9	1,097.36399	9	1,162.76112	9	1,228.15826	9	1,293.55540
790	1,033.27479	840	1,098.67193	890	1,164.06907	940	1,229.46621	990	1,294.86334
1	1,034.58273	1	1,099.97987	1	1,165.37701	1	1,230.77415	1	1,296.17129
2	1,035.89068	2	1,101.28781	2	1,166.68495	2	1,232.08209	2	1,297.47923
3	1,037.19862	3	1,102.59576	3	1,167.99290	3	1,233.39003	3	1,298.78717
4	1,038.50656	4	1,103.90370	4	1,169.30084	4	1,234.69798	4	1,300.09512
5	1,039.81450	5	1,105.21164	5	1,170.60878	5	1,236.00592	5	1,301.40306
6	1,041.12245	6	1,106.51959	6	1,171.91672	6	1,237.31386	6	1,302.71100
7	1,042.43039	7	1,107.82753	7	1,173.22467	7	1,238.62181	7	1,304.01894
8	1,043.73833	8	1,109.13547	8	1,174.53261	8	1,239.92975	8	1,305.32689
9	1,045.04627	9	1,110.44341	9	1,175.84055	9	1,241.23769	9	1,306.63483

VOLUME—CUBIC YARDS TO CUBIC METRES

[Reduction factor: 1 cubic yard = 0.7645594453 cubic metre]

Cubic yards	Cubic metres	Cubic yards	Cubic metres	Cubic yards	Cubic metres	Cubic yards	Cubic metres	Cubic yards	Cubic metres
0		50	38.22797	100	76.45594	150	114.68392	200	152.91189
1	0.76456	1	38.99253	1	77.22050	1	115.44848	1	153.67645
2	1.52912	2	39.75709	2	77.98506	2	116.21304	2	154.44101
3	2.29368	3	40.52165	3	78.74962	3	116.97760	3	155.20557
4	3.05824	4	41.28621	4	79.51418	4	117.74215	4	155.97013
5	3.82280	5	42.05077	5	80.27874	5	118.50671	5	156.73469
6	4.58736	6	42.81533	6	81.04330	6	119.27127	6	157.49925
7	5.35192	7	43.57989	7	81.80786	7	120.03583	7	158.26381
8	6.11648	8	44.34445	8	82.57242	8	120.80039	8	159.02836
9	6.88104	9	45.10901	9	83.33698	9	121.56495	9	159.79292
10	7.64559	60	45.87357	110	84.10154	160	122.32951	210	160.55748
1	8.41015	1	46.63813	1	84.86610	1	123.09407	1	161.32204
2	9.17471	2	47.40269	2	85.63066	2	123.85863	2	162.08660
3	9.93927	3	48.16725	3	86.39522	3	124.62319	3	162.85116
4	10.70383	4	48.93180	4	87.15978	4	125.38775	4	163.61572
5	11.46839	5	49.69636	5	87.92434	5	126.15231	5	164.38028
6	12.23295	6	50.46092	6	88.68890	6	126.91687	6	165.14484
7	12.99751	7	51.22548	7	89.45346	7	127.68143	7	165.90940
8	13.76207	8	51.99004	8	90.21801	8	128.44599	8	166.67396
9	14.52663	9	52.75460	9	90.98257	9	129.21055	9	167.43852
20	15.29119	70	53.51916	120	91.74713	170	129.97511	220	168.20308
1	16.05575	1	54.28372	1	92.51169	1	130.73967	1	168.96764
2	16.82031	2	55.04828	2	93.27625	2	131.50422	2	169.73220
3	17.58487	3	55.81284	3	94.04081	3	132.26878	3	170.49676
4	18.34943	4	56.57740	4	94.80537	4	133.03334	4	171.26132
5	19.11399	5	57.34196	5	95.56993	5	133.79790	5	172.02588
6	19.87855	6	58.10652	6	96.33449	6	134.56246	6	172.79043
7	20.64311	7	58.87108	7	97.09905	7	135.32702	7	173.55499
8	21.40766	8	59.63564	8	97.86361	8	136.09158	8	174.31955
9	22.17222	9	60.40020	9	98.62817	9	136.85614	9	175.08411
30	22.93678	80	61.16476	130	99.39273	180	137.62070	230	175.84867
1	23.70134	1	61.92932	1	100.15729	1	138.38526	1	176.61323
2	24.46590	2	62.69387	2	100.92185	2	139.14982	2	177.37779
3	25.23046	3	63.45843	3	101.68641	3	139.91438	3	178.14235
4	25.99502	4	64.22299	4	102.45097	4	140.67894	4	178.90691
5	26.75958	5	64.98755	5	103.21553	5	141.44350	5	179.67147
6	27.52414	6	65.75211	6	103.98008	6	142.20806	6	180.43603
7	28.28870	7	66.51667	7	104.74464	7	142.97262	7	181.20059
8	29.05326	8	67.28123	8	105.50920	8	143.73718	8	181.96515
9	29.81782	9	68.04579	9	106.27376	9	144.50174	9	182.72971
40	30.58238	90	68.81035	140	107.03832	190	145.26629	240	183.49427
1	31.34694	1	69.57491	1	107.80288	1	146.03085	1	184.25883
2	32.11150	2	70.33947	2	108.56744	2	146.79541	2	185.02339
3	32.87606	3	71.10403	3	109.33200	3	147.55997	3	185.78795
4	33.64062	4	71.86859	4	110.09656	4	148.32453	4	186.55250
5	34.40518	5	72.63315	5	110.86112	5	149.08909	5	187.31706
6	35.16973	6	73.39771	6	111.62568	6	149.85365	6	188.08162
7	35.93429	7	74.16227	7	112.39024	7	150.61821	7	188.84618
8	36.69885	8	74.92683	8	113.15480	8	151.38277	8	189.61074
9	37.46341	9	75.69139	9	113.91936	9	152.14733	9	190.37530

VOLUME—CUBIC YARDS TO CUBIC METRES

[Reduction factor: 1 cubic yard = 0.7645594453 cubic metre]

Cubic yards	Cubic metres	Cubic yards	Cubic metres	Cubic yards	Cubic metres	Cubic yards	Cubic metres	Cubic yards	Cubic metres
250	191.13986	300	229.36783	350	267.59581	400	305.82378	450	344.05175
1	191.90442	1	230.13239	1	268.36037	1	306.58834	1	344.81631
2	192.66898	2	230.89695	2	269.12492	2	307.35290	2	345.58087
3	193.43354	3	231.66151	3	269.88948	3	308.11746	3	346.34543
4	194.19810	4	232.42607	4	270.65404	4	308.88202	4	347.10999
5	194.96266	5	233.19063	5	271.41860	5	309.64658	5	347.87455
6	195.72722	6	233.95519	6	272.18316	6	310.41113	6	348.63911
7	196.49178	7	234.71975	7	272.94772	7	311.17569	7	349.40367
8	197.25634	8	235.48431	8	273.71228	8	311.94025	8	350.16823
9	198.02090	9	236.24887	9	274.47684	9	312.70481	9	350.93279
260	198.78546	310	237.01343	360	275.24140	410	313.46937	460	351.69734
1	199.55002	1	237.77799	1	276.00596	1	314.23393	1	352.46190
2	200.31457	2	238.54255	2	276.77052	2	314.99849	2	353.22646
3	201.07913	3	239.30711	3	277.53508	3	315.76305	3	353.99102
4	201.84369	4	240.07167	4	278.29964	4	316.52761	4	354.75558
5	202.60825	5	240.83623	5	279.06420	5	317.29217	5	355.52014
6	203.37281	6	241.60078	6	279.82876	6	318.05673	6	356.28470
7	204.13737	7	242.36534	7	280.59332	7	318.82129	7	357.04926
8	204.90193	8	243.12990	8	281.35788	8	319.58585	8	357.81382
9	205.66649	9	243.89446	9	282.12244	9	320.35041	9	358.57838
270	206.43105	320	244.65902	370	282.88699	420	321.11497	470	359.34294
1	207.19561	1	245.42358	1	283.65155	1	321.87953	1	360.10750
2	207.96017	2	246.18814	2	284.41611	2	322.64409	2	360.87206
3	208.72473	3	246.95270	3	285.18067	3	323.40865	3	361.63662
4	209.48929	4	247.71726	4	285.94523	4	324.17320	4	362.40118
5	210.25385	5	248.48182	5	286.70979	5	324.93776	5	363.16574
6	211.01841	6	249.24638	6	287.47435	6	325.70232	6	363.93030
7	211.78297	7	250.01094	7	288.23891	7	326.46688	7	364.69486
8	212.54753	8	250.77550	8	289.00347	8	327.23144	8	365.45941
9	213.31209	9	251.54006	9	289.76803	9	327.99600	9	366.22397
280	214.07664	330	252.30462	380	290.53259	430	328.76056	480	366.98853
1	214.84120	1	253.06918	1	291.29715	1	329.52512	1	367.75309
2	215.60576	2	253.83374	2	292.06171	2	330.28968	2	368.51765
3	216.37032	3	254.59830	3	292.82627	3	331.05424	3	369.28221
4	217.13488	4	255.36285	4	293.59083	4	331.81880	4	370.04677
5	217.89944	5	256.12741	5	294.35539	5	332.58336	5	370.81133
6	218.66400	6	256.89197	6	295.11995	6	333.34792	6	371.57589
7	219.42856	7	257.65653	7	295.88451	7	334.11248	7	372.34045
8	220.19312	8	258.42109	8	296.64906	8	334.87704	8	373.10501
9	220.95768	9	259.18565	9	297.41362	9	335.64160	9	373.86957
290	221.72224	340	259.95021	390	298.17818	440	336.40616	490	374.63413
1	222.48680	1	260.71477	1	298.94274	1	337.17072	1	375.39869
2	223.25136	2	261.47933	2	299.70730	2	337.93527	2	376.16325
3	224.01592	3	262.24389	3	300.47186	3	338.69983	3	376.92781
4	224.78048	4	263.00845	4	301.23642	4	339.46439	4	377.69237
5	225.54504	5	263.77301	5	302.00098	5	340.22895	5	378.45693
6	226.30960	6	264.53757	6	302.76554	6	340.99351	6	379.22148
7	227.07416	7	265.30213	7	303.53010	7	341.75807	7	379.98604
8	227.83871	8	266.06669	8	304.29466	8	342.52263	8	380.75060
9	228.60327	9	266.83125	9	305.05922	9	343.28719	9	381.51516

VOLUME—CUBIC YARDS TO CUBIC METRES

[Reduction factor: 1 cubic yard = 0.7645594453 cubic metre]

Cubic yards	Cubic metres	Cubic yards	Cubic metres	Cubic yards	Cubic metres	Cubic yards	Cubic metres	Cubic yards	Cubic metres
500	382.27972	550	420.50769	600	458.73567	650	496.96364	700	535.19161
1	383.04428	1	421.27225	1	459.50023	1	497.72820	1	535.95617
2	383.80884	2	422.03681	2	460.26479	2	498.49276	2	536.72073
3	384.57340	3	422.80137	3	461.02935	3	499.25732	3	537.48529
4	385.33796	4	423.56593	4	461.79390	4	500.02188	4	538.24985
5	386.10252	5	424.33049	5	462.55846	5	500.78644	5	539.01441
6	386.86708	6	425.09505	6	463.32302	6	501.55100	6	539.77897
7	387.63164	7	425.85961	7	464.08758	7	502.31556	7	540.54353
8	388.39620	8	426.62417	8	464.85214	8	503.08012	8	541.30809
9	389.16076	9	427.38873	9	465.61670	9	503.84467	9	542.07265
510	389.92532	560	428.15329	610	466.38126	660	504.60923	710	542.83721
1	390.68988	1	428.91785	1	467.14582	1	505.37379	1	543.60177
2	391.45444	2	429.68241	2	467.91038	2	506.13835	2	544.36633
3	392.21900	3	430.44697	3	468.67494	3	506.90291	3	545.13088
4	392.98355	4	431.21153	4	469.43950	4	507.66747	4	545.89544
5	393.74811	5	431.97609	5	470.20406	5	508.43203	5	546.66000
6	394.51267	6	432.74065	6	470.96862	6	509.19659	6	547.42456
7	395.27723	7	433.50521	7	471.73318	7	509.96115	7	548.18912
8	396.04179	8	434.26976	8	472.49774	8	510.72571	8	548.95368
9	396.80635	9	435.03432	9	473.26230	9	511.49027	9	549.71824
520	397.57091	570	435.79888	620	474.02686	670	512.25483	720	550.48280
1	398.33547	1	436.56344	1	474.79142	1	513.01939	1	551.24736
2	399.10003	2	437.32800	2	475.55597	2	513.78395	2	552.01192
3	399.86459	3	438.09256	3	476.32053	3	514.54851	3	552.77648
4	400.62915	4	438.85712	4	477.08509	4	515.31307	4	553.54104
5	401.39371	5	439.62168	5	477.84965	5	516.07763	5	554.30560
6	402.15827	6	440.38624	6	478.61421	6	516.84219	6	555.07016
7	402.92283	7	441.15080	7	479.37877	7	517.60674	7	555.83472
8	403.68739	8	441.91536	8	480.14333	8	518.37130	8	556.59928
9	404.45195	9	442.67992	9	480.90789	9	519.13586	9	557.36384
530	405.21651	580	443.44448	630	481.67245	680	519.90042	730	558.12840
1	405.98107	1	444.20904	1	482.43701	1	520.66498	1	558.89295
2	406.74562	2	444.97360	2	483.20157	2	521.42954	2	559.65751
3	407.51018	3	445.73816	3	483.96613	3	522.19410	3	560.42207
4	408.27474	4	446.50272	4	484.73069	4	522.95866	4	561.18663
5	409.03930	5	447.26728	5	485.49525	5	523.72322	5	561.95119
6	409.80386	6	448.03183	6	486.25981	6	524.40770	6	562.71575
7	410.56842	7	448.79639	7	487.02437	7	525.25234	7	563.48031
8	411.33298	8	449.56095	8	487.78893	8	526.01690	8	564.24487
9	412.09754	9	450.32551	9	488.55349	9	526.78146	9	565.00943
540	412.86210	590	451.09007	640	489.31804	690	527.54602	740	565.77399
1	413.62666	1	451.85463	1	490.08260	1	528.31058	1	566.53855
2	414.39122	2	452.61919	2	490.84716	2	529.07514	2	567.30311
3	415.15578	3	453.38375	3	491.61172	3	529.83970	3	568.06767
4	415.92034	4	454.14831	4	492.37628	4	530.60426	4	568.83223
5	416.68490	5	454.91287	5	493.14084	5	531.36881	5	569.59679
6	417.44946	6	455.67743	6	493.90540	6	532.13337	6	570.36135
7	418.21402	7	456.44199	7	494.66996	7	532.89793	7	571.12591
8	418.97858	8	457.20655	8	495.43452	8	533.66249	8	571.89047
9	419.74314	9	457.97111	9	496.19908	9	534.42705	9	572.65502

VOLUME—CUBIC YARDS TO CUBIC METRES

[Reduction factor: 1 cubic yard = 0.7645594453 cubic metre]

Cubic yards	Cubic metres	Cubic yards	Cubic metres	Cubic yards	Cubic metres	Cubic yards	Cubic metres	Cubic yards	Cubic metres
750	573.41958	800	611.64756	850	649.87553	900	688.10350	950	726.33147
1	574.18414	1	612.41212	1	650.64009	1	688.86806	1	727.09603
2	574.94870	2	613.17668	2	651.40465	2	689.63262	2	727.86059
3	575.71326	3	613.94123	3	652.16921	3	690.39718	3	728.62515
4	576.47782	4	614.70579	4	652.93377	4	691.16174	4	729.38971
5	577.24238	5	615.47035	5	653.69833	5	691.92630	5	730.15427
6	578.00694	6	616.23491	6	654.46289	6	692.69086	6	730.91883
7	578.77150	7	616.99947	7	655.22744	7	693.45542	7	731.68339
8	579.53606	8	617.76403	8	655.99200	8	694.21998	8	732.44795
9	580.30062	9	618.52859	9	656.75656	9	694.98454	9	733.21251
760	581.06518	810	619.29315	860	657.52112	910	695.74910	960	733.97707
1	581.82974	1	620.05771	1	658.28568	1	696.51365	1	734.74163
2	582.59430	2	620.82227	2	659.05024	2	697.27821	2	735.50619
3	583.35886	3	621.58683	3	659.81480	3	698.04277	3	736.27075
4	584.12342	4	622.35139	4	660.57936	4	698.80733	4	737.03531
5	584.88798	5	623.11595	5	661.34392	5	699.57189	5	737.79986
6	585.65254	6	623.88051	6	662.10848	6	700.33645	6	738.56442
7	586.41709	7	624.64507	7	662.87304	7	701.10101	7	739.32898
8	587.18165	8	625.40963	8	663.63760	8	701.86557	8	740.09354
9	587.94621	9	626.17419	9	664.40216	9	702.63013	9	740.85810
770	588.71077	820	626.93875	870	665.16672	920	703.39469	970	741.62266
1	589.47533	1	627.70330	1	665.93128	1	704.15925	1	742.38722
2	590.23989	2	628.46786	2	666.69584	2	704.92381	2	743.15178
3	591.00445	3	629.23242	3	667.46040	3	705.68837	3	743.91634
4	591.76901	4	629.99698	4	668.22496	4	706.45293	4	744.68090
5	592.53357	5	630.76154	5	668.98951	5	707.21749	5	745.44546
6	593.29813	6	631.52610	6	669.75407	6	707.98205	6	746.21002
7	594.06269	7	632.29066	7	670.51863	7	708.74661	7	746.97458
8	594.82725	8	633.05522	8	671.28319	8	709.51117	8	747.73914
9	595.59181	9	633.81978	9	672.04775	9	710.27572	9	748.50370
780	596.35637	830	634.58434	880	672.81231	930	711.04028	980	749.26826
1	597.12093	1	635.34890	1	673.57687	1	711.80484	1	750.03282
2	597.88549	2	636.11346	2	674.34143	2	712.56940	2	750.79738
3	598.65005	3	636.87802	3	675.10599	3	713.33396	3	751.56193
4	599.41461	4	637.64258	4	675.87055	4	714.09852	4	752.32649
5	600.17916	5	638.40714	5	676.63511	5	714.86308	5	753.09105
6	600.94372	6	639.17170	6	677.39967	6	715.62764	6	753.85561
7	601.70828	7	639.93626	7	678.16423	7	716.39220	7	754.62017
8	602.47284	8	640.70082	8	678.92879	8	717.15676	8	755.38473
9	603.23740	9	641.46537	9	679.69335	9	717.92132	9	756.14929
790	604.00196	840	642.22993	890	680.45791	940	718.68588	990	756.91385
1	604.76652	1	642.99449	1	681.22247	1	719.45044	1	757.67841
2	605.53108	2	643.75905	2	681.98703	2	720.21500	2	758.44297
3	606.29564	3	644.52361	3	682.75158	3	720.97956	3	759.20753
4	607.06020	4	645.28817	4	683.51614	4	721.74412	4	759.97209
5	607.82476	5	646.05273	5	684.28070	5	722.50868	5	760.73665
6	608.58932	6	646.81729	6	685.04526	6	723.27324	6	761.50121
7	609.35388	7	647.58185	7	685.80982	7	724.03779	7	762.26577
8	610.11844	8	648.34641	8	686.57438	8	724.80235	8	763.03033
9	610.88300	9	649.11097	9	687.33894	9	725.56691	9	763.79489

CAPACITY—LITRES TO LIQUID QUARTS

[Reduction factor: 1 litre = 1.0567104 quarts]

Litres	Liquid quarts	Litres	Liquid quarts	Litres	Liquid quarts	Litres	Liquid quarts	Litres	Liquid quarts
0		50	52.8355	100	105.671	150	158.507	200	211.342
1	1.0567	1	53.8922	1	106.728	1	159.563	1	212.399
2	2.1134	2	54.9489	2	107.784	2	160.620	2	213.456
3	3.1701	3	56.0057	3	108.841	3	161.677	3	214.512
4	4.2268	4	57.062+	4	109.898	4	162.733	4	215.569
5	5.2836	5	58.1191	5	110.955	5	163.790	5	216.626
6	6.3403	6	59.1758	6	112.011	6	164.847	6	217.682
7	7.3970	7	60.2325	7	113.068	7	165.904	7	218.739
8	8.4537	8	61.2892	8	114.125	8	166.960	8	219.796
9	9.5104	9	62.3459	9	115.181	9	168.017	9	220.852
10	10.5671	60	63.4026	110	116.238	160	169.074	210	221.909
1	11.6238	1	64.4593	1	117.295	1	170.130	1	222.966
2	12.6805	2	65.5160	2	118.352	2	171.187	2	224.023
3	13.7372	3	66.5728	3	119.408	3	172.244	3	225.079
4	14.7939	4	67.6295	4	120.465	4	173.301	4	226.136
5	15.8507	5	68.6862	5	121.522	5	174.357	5	227.193
6	16.9074	6	69.7429	6	122.578	6	175.414	6	228.249
7	17.9641	7	70.7996	7	123.635	7	176.471	7	229.306
8	19.0208	8	71.8563	8	124.692	8	177.527	8	230.363
9	20.0775	9	72.9130	9	125.749	9	178.584	9	231.420
20	21.1342	70	73.9697	120	126.805	170	179.641	220	232.476
1	22.1909	1	75.0264	1	127.862	1	180.697	1	233.533
2	23.2476	2	76.0831	2	128.919	2	181.754	2	234.590
3	24.3043	3	77.1399	3	129.975	3	182.811	3	235.646
4	25.3610	4	78.1966	4	131.032	4	183.868	4	236.703
5	26.4178	5	79.2533	5	132.089	5	184.924	5	237.760
6	27.4745	6	80.3100	6	133.146	6	185.981	6	238.817
7	28.5312	7	81.3667	7	134.202	7	187.038	7	239.873
8	29.5879	8	82.4234	8	135.259	8	188.094	8	240.930
9	30.6446	9	83.4801	9	136.316	9	189.151	9	241.987
30	31.7013	80	84.5368	130	137.372	180	190.208	230	243.043
1	32.7580	1	85.5935	1	138.429	1	191.265	1	244.100
2	33.8147	2	86.6503	2	139.486	2	192.321	2	245.157
3	34.8714	3	87.7070	3	140.542	3	193.378	3	246.214
4	35.9282	4	88.7637	4	141.599	4	194.435	4	247.270
5	36.9849	5	89.8204	5	142.656	5	195.491	5	248.327
6	38.0416	6	90.8771	6	143.713	6	196.548	6	249.384
7	39.0983	7	91.9338	7	144.769	7	197.605	7	250.440
8	40.1550	8	92.9905	8	145.826	8	198.662	8	251.497
9	41.2117	9	94.0472	9	146.883	9	199.718	9	252.554
40	42.2684	90	95.1039	140	147.939	190	200.775	240	253.610
1	43.3251	1	96.1606	1	148.996	1	201.832	1	254.667
2	44.3818	2	97.2174	2	150.053	2	202.388	2	255.724
3	45.4385	3	98.2741	3	151.110	3	203.945	3	256.781
4	46.4953	4	99.3308	4	152.166	4	205.002	4	257.837
5	47.5520	5	100.3875	5	153.223	5	206.059	5	258.894
6	48.6087	6	101.4442	6	154.280	6	207.115	6	259.951
7	49.6654	7	102.5009	7	155.336	7	208.172	7	261.007
8	50.7221	8	103.5576	8	156.393	8	209.229	8	262.064
9	51.7788	9	104.6143	9	157.450	9	210.285	9	263.121

CAPACITY—LITRES TO LIQUID QUARTS

[Reduction factor: 1 litre = 1.0567104 quarts]

Litres	Liquid quarts	Litres	Liquid quarts	Litres	Liquid quarts	Litres	Liquid quarts	Litres	Liquid quarts
250	264.178	300	317.013	350	369.849	400	422.684	450	475.520
1	265.234	1	318.070	1	370.905	1	423.741	1	476.576
2	266.291	2	319.127	2	371.962	2	424.798	2	477.633
3	267.348	3	320.183	3	373.019	3	425.854	3	478.690
4	268.404	4	321.240	4	374.075	4	426.911	4	479.747
5	269.461	5	322.297	5	375.132	5	427.968	5	480.803
6	270.518	6	323.353	6	376.189	6	429.024	6	481.860
7	271.575	7	324.410	7	377.246	7	430.081	7	482.917
8	272.631	8	325.467	8	378.302	8	431.138	8	483.973
9	273.688	9	326.524	9	379.359	9	432.195	9	485.030
260	274.745	310	327.580	360	380.416	410	433.251	460	486.087
1	275.801	1	328.637	1	381.472	1	434.308	1	487.143
2	276.858	2	329.694	2	382.529	2	435.365	2	488.200
3	277.915	3	330.750	3	383.586	3	736.421	3	489.257
4	278.972	4	331.807	4	384.643	4	437.478	4	490.314
5	280.028	5	332.864	5	385.699	5	438.535	5	491.370
6	281.085	6	333.920	6	386.756	6	439.592	6	492.427
7	282.142	7	334.977	7	387.813	7	440.648	7	493.484
8	283.198	8	336.034	8	388.869	8	441.705	8	494.540
9	284.255	9	337.091	9	389.926	9	442.762	9	495.597
270	285.312	320	338.147	370	390.983	420	443.818	470	496.654
1	286.369	1	339.204	1	392.040	1	444.875	1	497.711
2	287.425	2	340.261	2	393.096	2	445.932	2	498.767
3	288.482	3	341.317	3	394.153	3	446.988	3	499.824
4	289.539	4	342.374	4	395.210	4	448.045	4	500.881
5	290.595	5	343.431	5	396.266	5	449.102	5	501.937
6	291.652	6	344.488	6	397.323	6	450.159	6	502.994
7	292.709	7	345.544	7	398.380	7	451.215	7	504.051
8	293.765	8	346.601	8	399.437	8	452.272	8	505.108
9	294.822	9	347.658	9	400.493	9	453.329	9	506.164
280	295.879	330	348.714	380	401.550	430	454.385	480	507.221
1	296.936	1	349.771	1	402.607	1	455.442	1	508.278
2	297.992	2	350.828	2	403.663	2	456.499	2	509.334
3	299.049	3	351.885	3	404.720	3	457.556	3	510.391
4	300.106	4	352.941	4	405.777	4	458.612	4	511.448
5	301.162	5	353.998	5	406.834	5	459.669	5	512.505
6	302.219	6	355.055	6	407.890	6	460.726	6	513.561
7	303.276	7	356.111	7	408.947	7	461.782	7	514.618
8	304.333	8	357.168	8	410.004	8	462.839	8	515.675
9	305.389	9	358.225	9	411.060	9	463.896	9	516.731
290	306.446	340	359.282	390	412.117	440	464.953	490	517.788
1	307.503	1	360.338	1	413.174	1	466.009	1	518.845
2	308.559	2	361.395	2	414.230	2	467.066	2	519.902
3	309.616	3	362.452	3	415.287	3	468.123	3	520.958
4	310.673	4	363.508	4	416.344	4	469.179	4	522.015
5	311.730	5	364.565	5	417.401	5	470.236	5	523.072
6	312.786	6	365.622	6	418.457	6	471.293	6	524.128
7	313.843	7	366.679	7	419.514	7	472.350	7	525.185
8	314.900	8	367.735	8	420.571	8	473.406	8	526.242
9	315.956	9	368.792	9	421.627	9	474.463	9	527.298

CAPACITY—LITRES TO LIQUID QUARTS

[Reduction factor: 1 litre = 1.0567104 quarts]

Litres	Liquid quarts	Litres	Liquid quarts	Litres	Liquid quarts	Litres	Liquid quarts	Litres	Liquid quarts
500	528. 355	550	581. 191	600	634. 026	650	686. 862	700	739. 697
1	529. 412	1	582. 247	1	635. 083	1	687. 918	1	740 754
2	530. 469	2	583. 304	2	636. 140	2	688. 975	2	741. 811
3	531. 525	3	584. 361	3	637. 196	3	690. 032	3	742. 867
4	532. 582	4	585. 418	4	638. 253	4	691. 089	4	743. 924
5	533. 639	5	586. 474	5	639. 310	5	692. 145	5	744. 981
6	534. 695	6	587. 531	6	640. 367	6	693. 202	6	746. 038
7	535. 752	7	588. 588	7	641. 423	7	694. 259	7	747. 094
8	536. 809	8	589. 644	8	642. 480	8	695. 315	8	748. 151
9	537. 866	9	590. 701	9	643. 537	9	696. 372	9	749. 208
510	538. 922	560	591. 758	610	644. 593	660	697. 429	710	750. 264
1	539. 979	1	592. 815	1	645. 650	1	698. 486	1	751. 321
2	541. 036	2	593. 871	2	646. 707	2	699. 542	2	752. 378
3	542. 092	3	594. 928	3	647. 763	3	700. 599	3	753. 435
4	543. 149	4	595. 985	4	648. 820	4	701. 656	4	754. 491
5	544. 206	5	597. 041	5	649. 877	5	702. 712	5	755. 548
6	545. 263	6	598. 098	6	650. 934	6	703. 769	6	756. 605
7	546. 319	7	599. 155	7	651. 990	7	704. 826	7	757. 661
8	547. 376	8	600. 212	8	653. 047	8	705. 883	8	758. 718
9	548. 433	9	601. 268	9	654. 104	9	706. 939	9	759. 775
520	549. 489	570	602. 325	620	655. 160	670	707. 996	720	760. 831
1	550. 546	1	603. 382	1	656. 217	1	709. 053	1	761. 888
2	551. 603	2	604. 438	2	657. 274	2	710. 109	2	762. 945
3	552. 660	3	605. 495	3	658. 331	3	711. 166	3	764. 002
4	553. 716	4	606. 552	4	659. 387	4	712. 223	4	765. 058
5	554. 773	5	607. 608	5	660. 444	5	713. 280	5	766. 115
6	555. 830	6	608. 665	6	661. 501	6	714. 336	6	767. 172
7	556. 886	7	609. 722	7	662. 557	7	715. 393	7	768. 228
8	557. 943	8	610. 779	8	663. 614	8	716. 450	8	769. 285
9	559. 000	9	611. 835	9	664. 671	9	717. 506	9	770. 342
530	560. 057	580	612. 892	630	665. 728	680	718. 563	730	771. 399
1	561. 113	1	613. 949	1	666. 784	1	719. 620	1	772. 455
2	562. 170	2	615. 005	2	667. 841	2	720. 676	2	773. 512
3	563. 227	3	616. 062	3	668. 898	3	721. 733	3	774. 569
4	564. 283	4	617. 119	4	669. 954	4	722. 790	4	775. 625
5	565. 340	5	618. 176	5	671. 011	5	723. 847	5	776. 682
6	566. 397	6	619. 232	6	672. 068	6	724. 903	6	777. 739
7	567. 453	7	620. 289	7	673. 125	7	725. 960	7	778. 796
8	568. 510	8	621. 346	8	674. 181	8	727. 017	8	779. 852
9	569. 567	9	622. 402	9	675. 238	9	728. 073	9	780. 909
540	570. 624	590	623. 459	640	676. 295	690	729. 130	740	781. 966
1	571. 680	1	624. 516	1	677. 351	1	730. 187	1	783. 022
2	572. 737	2	625. 573	2	678. 408	2	731. 244	2	784. 079
3	573. 794	3	626. 629	3	679. 465	3	732. 300	3	785. 136
4	574. 850	4	627. 686	4	680. 521	4	733. 357	4	786. 193
5	575. 907	5	628. 743	5	681. 578	5	734. 414	5	787. 249
6	576. 964	6	629. 799	6	682. 635	6	735. 470	6	788. 306
7	578. 021	7	630. 856	7	683. 692	7	736. 527	7	789. 363
8	579. 077	8	631. 913	8	684. 748	8	737. 584	8	790. 419
9	580. 134	9	632. 970	9	685. 805	9	738. 641	9	791. 476

CAPACITY—LITRES TO LIQUID QUARTS

[Reduction factor: 1 litre = 1.0567104 quarts]

Litres	Liquid quarts	Litres	Liquid quarts	Litres	Liquid quarts	Litres	Liquid quarts	Litres	Liquid quarts
750	792. 533	800	845. 368	850	898. 204	900	951. 039	950	1,003. 875
1	793. 590	1	846. 425	1	899. 261	1	952. 096	1	1,004. 932
2	794. 646	2	847. 482	2	900. 317	2	953. 153	2	1,005. 988
3	795. 703	3	848. 538	3	901. 374	3	954. 209	3	1,007. 045
4	796. 760	4	849. 595	4	902. 431	4	955. 266	4	1,008. 102
5	797. 816	5	850. 652	5	903. 487	5	956. 323	5	1,009. 158
6	798. 873	6	851. 709	6	904. 544	6	957. 380	6	1,010. 215
7	799. 930	7	852. 765	7	905. 601	7	958. 436	7	1,011. 272
8	800. 986	8	853. 822	8	906. 658	8	959. 493	8	1,012. 329
9	802. 043	9	854. 879	9	907. 714	9	960. 550	9	1,013. 385
760	803. 100	810	855. 935	860	908. 771	910	961. 606	960	1,014. 442
1	804. 157	1	856. 992	1	909. 828	1	962. 663	1	1,015. 499
2	805. 213	2	858. 049	2	910. 884	2	963. 720	2	1,016. 555
3	806. 270	3	859. 106	3	911. 941	3	964. 777	3	1,017. 612
4	807. 327	4	860. 162	4	912. 998	4	965. 833	4	1,018. 669
5	808. 383	5	861. 219	5	914. 054	5	966. 890	5	1,019. 726
6	809. 440	6	862. 276	6	915. 111	6	967. 947	6	1,020. 782
7	810. 497	7	863. 332	7	916. 168	7	969. 003	7	1,021. 839
8	811. 554	8	864. 389	8	917. 225	8	970. 060	8	1,022. 896
9	812. 610	9	865. 446	9	918. 281	9	971. 117	9	1,023. 952
770	813. 667	820	866. 503	870	919. 338	920	972. 174	970	1,025. 009
1	814. 724	1	867. 559	1	920. 395	1	973. 230	1	1,026. 066
2	815. 780	2	868. 616	2	921. 451	2	974. 287	2	1,027. 123
3	816. 837	3	869. 673	3	922. 508	3	975. 344	3	1,028. 179
4	817. 894	4	870. 729	4	923. 565	4	976. 400	4	1,029. 236
5	818. 951	5	871. 786	5	924. 622	5	977. 457	5	1,030. 293
6	820. 007	6	872. 843	6	925. 678	6	978. 514	6	1,031. 349
7	821. 064	7	873. 900	7	926. 735	7	979. 571	7	1,032. 406
8	822. 121	8	874. 956	8	927. 792	8	980. 627	8	1,033. 463
9	823. 177	9	876. 013	9	928. 848	9	981. 684	9	1,034. 519
780	824. 234	830	877. 070	880	929. 905	930	982. 741	980	1,035. 576
1	825. 291	1	878. 126	1	930. 962	1	983. 797	1	1,036. 633
2	826. 348	2	879. 183	2	932. 019	2	984. 854	2	1,037. 690
3	827. 404	3	880. 240	3	933. 075	3	985. 911	3	1,038. 746
4	828. 461	4	881. 296	4	934. 132	4	986. 968	4	1,039. 803
5	829. 518	5	882. 353	5	935. 189	5	988. 024	5	1,040. 860
6	830. 574	6	883. 410	6	936. 245	6	989. 081	6	1,041. 916
7	831. 631	7	884. 467	7	937. 302	7	990. 138	7	1,042. 973
8	832. 688	8	885. 523	8	938. 359	8	991. 194	8	1,044. 030
9	833. 745	9	886. 580	9	939. 416	9	992. 251	9	1,045. 087
790	834. 801	840	887. 637	890	940. 472	940	993. 308	990	1,046. 143
1	835. 858	1	888. 693	1	941. 529	1	994. 364	1	1,047. 200
2	836. 915	2	889. 750	2	942. 586	2	995. 421	2	1,048. 257
3	837. 971	3	890. 807	3	943. 642	3	996. 478	3	1,049. 313
4	839. 028	4	891. 864	4	944. 699	4	997. 535	4	1,050. 370
5	840. 085	5	892. 920	5	945. 756	5	998. 591	5	1,051. 427
6	841. 141	6	893. 977	6	946. 813	6	999. 648	6	1,052. 484
7	842. 198	7	895. 034	7	947. 869	7	1,000. 705	7	1,053. 540
8	843. 255	8	896. 090	8	948. 926	8	1,001. 761	8	1,054. 597
9	844. 312	9	897. 147	9	949. 983	9	1,002. 818	9	1,055. 654

CAPACITY—LIQUID QUARTS TO LITRES

[Reduction factor: 1 liquid quart = 0.94633307 litre]

Liquid quarts	Litres	Liquid quarts	Litres	Liquid quarts	Litres	Liquid quarts	Litres	Liquid quarts	Litres
0		50	47.3167	100	94.633	150	141.950	200	189.267
1	0.9463	1	48.2630	1	95.580	1	142.896	1	190.213
2	1.8927	2	49.2093	2	96.526	2	143.843	2	191.159
3	2.8390	3	50.1557	3	97.472	3	144.789	3	192.106
4	3.7853	4	51.1020	4	98.419	4	145.735	4	193.052
5	4.7317	5	52.0483	5	99.365	5	146.682	5	193.998
6	5.6780	6	52.9947	6	100.311	6	147.628	6	194.945
7	6.6243	7	53.9410	7	101.258	7	148.574	7	195.891
8	7.5707	8	54.8873	8	102.204	8	149.521	8	196.837
9	8.5170	9	55.8337	9	103.150	9	150.467	9	197.784
10	9.4633	60	56.7800	110	104.097	160	151.413	210	198.730
1	10.4097	1	57.7263	1	105.043	1	152.360	1	199.676
2	11.3560	2	58.6727	2	105.989	2	153.306	2	200.623
3	12.3023	3	59.6190	3	106.936	3	154.252	3	201.569
4	13.2487	4	60.5653	4	107.882	4	155.199	4	202.515
5	14.1950	5	61.5116	5	108.828	5	156.145	5	203.462
6	15.1413	6	62.4580	6	109.775	6	157.091	6	204.408
7	16.0877	7	63.4043	7	110.721	7	158.038	7	205.354
8	17.0340	8	64.3506	8	111.667	8	158.984	8	206.301
9	17.9803	9	65.2970	9	112.614	9	159.930	9	207.247
20	18.9267	70	66.2433	120	113.560	170	160.877	220	208.193
1	19.8730	1	67.1896	1	114.506	1	161.823	1	209.140
2	20.8193	2	68.1360	2	115.453	2	162.769	2	210.086
3	21.7657	3	69.0823	3	116.399	3	163.716	3	211.032
4	22.7120	4	70.0286	4	117.345	4	164.662	4	211.979
5	23.6583	5	70.9750	5	118.292	5	165.608	5	212.925
6	24.6047	6	71.9213	6	119.238	6	166.555	6	213.871
7	25.5510	7	72.8676	7	120.184	7	167.501	7	214.818
8	26.4973	8	73.8140	8	121.131	8	168.447	8	215.764
9	27.4437	9	74.7603	9	122.077	9	169.394	9	216.710
30	28.3900	80	75.7066	130	123.023	180	170.340	230	217.657
1	29.3363	1	76.6530	1	123.970	1	171.286	1	218.603
2	30.2827	2	77.5993	2	124.916	2	172.233	2	219.549
3	31.2290	3	78.5456	3	125.862	3	173.179	3	220.496
4	32.1753	4	79.4920	4	126.809	4	174.125	4	221.442
5	33.1217	5	80.4383	5	127.755	5	175.072	5	223.388
6	34.0680	6	81.3846	6	128.701	6	176.018	6	223.335
7	35.0143	7	82.3310	7	129.648	7	176.964	7	224.281
8	35.9607	8	83.2773	8	130.594	8	177.911	8	225.227
9	36.9070	9	84.2236	9	131.540	9	178.857	9	226.174
40	37.8533	90	85.1700	140	132.487	190	179.803	240	227.120
1	38.7997	1	86.1163	1	133.433	1	180.750	1	228.066
2	39.7460	2	87.0626	2	134.379	2	181.696	2	229.013
3	40.6923	3	88.0090	3	135.326	3	182.642	3	229.959
4	41.6387	4	88.9553	4	136.272	4	183.589	4	230.905
5	42.5850	5	89.9016	5	137.218	5	184.535	5	231.852
6	43.5313	6	90.8480	6	138.165	6	185.481	6	232.798
7	44.4777	7	91.7943	7	139.111	7	186.428	7	233.744
8	45.4240	8	92.7406	8	140.057	8	187.374	8	234.691
9	46.3703	9	93.6870	9	141.004	9	188.320	9	235.637

CAPACITY—LIQUID QUARTS TO LITRES

[Reduction factor: 1 liquid quart = 0.94633307 litre]

Liquid quarts	Litres	Liquid quarts	Litres	Liquid quarts	Litres	Liquid quarts	Litres	Liquid quarts	Litres
250	236.583	300	283.900	350	331.217	400	378.533	450	425.850
1	237.530	1	284.846	1	332.163	1	379.480	1	426.796
2	238.476	2	285.793	2	333.109	2	380.426	2	427.743
3	239.422	3	286.739	3	334.056	3	381.372	3	428.689
4	240.369	4	287.685	4	335.002	4	382.319	4	429.635
5	241.315	5	288.632	5	335.948	5	383.265	5	430.582
6	242.261	6	289.578	6	336.895	6	384.211	6	431.528
7	243.208	7	290.524	7	337.841	7	385.158	7	432.474
8	244.154	8	291.471	8	338.787	8	386.104	8	433.421
9	245.100	9	292.417	9	339.734	9	387.050	9	434.367
260	246.047	310	293.363	360	340.680	410	387.997	460	435.313
1	246.993	1	294.310	1	341.626	1	388.943	1	436.260
2	247.939	2	295.256	2	342.573	2	389.889	2	437.206
3	248.886	3	296.202	3	343.519	3	390.836	3	438.152
4	249.832	4	297.149	4	344.465	4	391.782	4	439.099
5	250.778	5	298.095	5	345.412	5	392.728	5	440.045
6	251.725	6	299.041	6	346.358	6	393.675	6	440.991
7	252.671	7	299.988	7	347.304	7	394.621	7	441.938
8	253.617	8	300.934	8	348.251	8	395.567	8	442.884
9	254.564	9	301.880	9	349.197	9	396.514	9	443.830
270	255.510	320	302.827	370	350.143	420	397.460	470	444.777
1	256.456	1	303.773	1	351.090	1	398.406	1	445.723
2	257.403	2	304.719	2	352.036	2	399.353	2	446.669
3	258.349	3	305.666	3	352.982	3	400.299	3	447.616
4	259.295	4	306.612	4	353.929	4	401.245	4	448.562
5	260.242	5	307.558	5	354.875	5	402.192	5	449.508
6	261.188	6	308.505	6	355.821	6	403.138	6	450.455
7	262.134	7	309.451	7	356.768	7	404.084	7	451.401
8	263.081	8	310.397	8	357.714	8	405.031	8	452.347
9	264.027	9	311.344	9	358.660	9	405.977	9	453.294
280	264.973	330	312.290	380	359.607	430	406.923	480	454.240
1	265.920	1	313.236	1	360.553	1	407.870	1	455.186
2	266.866	2	314.183	2	361.499	2	408.816	2	456.133
3	267.812	3	315.129	3	362.446	3	409.762	3	457.079
4	268.759	4	316.075	4	363.392	4	410.709	4	458.025
5	269.705	5	317.022	5	364.338	5	411.655	5	458.972
6	270.651	6	317.968	6	365.285	6	412.601	6	459.918
7	271.598	7	318.914	7	366.231	7	413.548	7	460.864
8	272.544	8	319.861	8	367.177	8	414.494	8	461.811
9	273.490	9	320.807	9	368.124	9	415.440	9	462.757
290	274.437	340	321.753	390	369.070	440	416.387	490	463.703
1	275.383	1	322.700	1	370.016	1	417.333	1	464.650
2	276.329	2	323.646	2	370.963	2	418.279	2	465.596
3	277.276	3	324.592	3	371.909	3	419.226	3	466.542
4	278.222	4	325.539	4	372.855	4	420.172	4	467.489
5	279.168	5	326.485	5	373.802	5	421.118	5	468.435
6	280.115	6	327.431	6	374.748	6	422.065	6	469.381
7	281.061	7	328.378	7	375.694	7	423.011	7	470.328
8	282.007	8	329.324	8	376.641	8	423.957	8	471.274
9	282.954	9	330.270	9	377.587	9	424.904	9	472.220

CAPACITY—LIQUID QUARTS TO LITRES

[Reduction factor: 1 liquid quart = 0.94633307 litre]

Liquid quarts	Litres	Liquid quarts	Litres	Liquid quarts	Litres	Liquid quarts	Litres	Liquid quarts	Litres
500	473.167	550	520.483	600	567.800	650	615.116	700	662.433
1	474.113	1	521.430	1	568.746	1	616.063	1	663.379
2	475.059	2	522.376	2	569.693	2	617.009	2	664.326
3	476.006	3	523.322	3	570.639	3	617.955	3	665.272
4	476.952	4	524.269	4	571.585	4	618.902	4	666.218
5	477.898	5	525.215	5	572.532	5	619.848	5	667.165
6	478.845	6	526.161	6	573.478	6	620.794	6	668.111
7	479.791	7	527.108	7	574.424	7	621.741	7	669.057
8	480.737	8	528.054	8	575.371	8	622.687	8	670.004
9	481.684	9	529.000	9	576.317	9	623.633	9	670.950
510	482.630	560	529.947	610	577.263	660	624.580	710	671.896
1	483.576	1	530.893	1	578.210	1	625.526	1	672.843
2	484.523	2	531.839	2	579.156	2	626.472	2	673.789
3	485.469	3	532.786	3	580.102	3	627.419	3	674.735
4	486.415	4	533.732	4	581.049	4	628.365	4	675.682
5	487.362	5	534.678	5	581.995	5	629.311	5	676.628
6	488.308	6	535.625	6	582.941	6	630.258	6	677.574
7	489.254	7	536.571	7	583.888	7	631.204	7	678.521
8	490.201	8	537.517	8	584.834	8	632.150	8	679.467
9	491.147	9	538.464	9	585.780	9	633.097	9	680.413
520	492.093	570	539.410	620	586.727	670	634.043	720	681.360
1	493.040	1	540.356	1	587.673	1	634.989	1	682.306
2	493.986	2	541.303	2	588.619	2	635.936	2	683.252
3	494.932	3	542.249	3	589.566	3	636.882	3	684.199
4	495.879	4	543.195	4	590.512	4	637.828	4	685.145
5	496.825	5	544.142	5	591.458	5	638.775	5	686.091
6	497.771	6	545.088	6	592.405	6	639.721	6	687.038
7	498.718	7	546.034	7	593.351	7	640.667	7	687.984
8	499.664	8	546.981	8	594.297	8	641.614	8	688.930
9	500.610	9	547.927	9	595.244	9	642.560	9	689.877
530	501.557	580	548.873	630	596.190	680	643.506	730	690.823
1	502.503	1	549.820	1	597.136	1	644.453	1	691.769
2	503.449	2	550.766	2	598.083	2	645.399	2	692.716
3	504.396	3	551.712	3	599.029	3	646.345	3	693.662
4	505.342	4	552.659	4	599.975	4	647.292	4	694.608
5	506.288	5	553.605	5	600.922	5	648.238	5	695.555
6	507.235	6	554.551	6	601.868	6	649.184	6	696.501
7	508.181	7	555.498	7	602.814	7	650.131	7	697.447
8	509.127	8	556.444	8	603.761	8	651.077	8	698.394
9	510.074	9	557.390	9	604.707	9	652.023	9	699.340
540	511.020	590	558.337	640	605.653	690	652.970	740	700.286
1	511.966	1	559.283	1	606.599	1	653.916	1	701.233
2	512.913	2	560.229	2	607.546	2	654.862	2	702.179
3	513.859	3	561.176	3	608.492	3	655.809	3	703.125
4	514.805	4	562.122	4	609.438	4	656.755	4	704.072
5	515.752	5	563.068	5	610.385	5	657.701	5	705.018
6	516.698	6	564.015	6	611.331	6	658.648	6	705.964
7	517.644	7	564.961	7	612.277	7	659.594	7	706.911
8	518.591	8	565.907	8	613.224	8	660.540	8	707.857
9	519.537	9	566.854	9	614.170	9	661.487	9	708.803

CAPACITY—LIQUID QUARTS TO LITRES

[Reduction factor: 1 liquid quart = 0.94633307 litre]

Liquid quarts	Litres	Liquid quarts	Litres	Liquid quarts	Litres	Liquid quarts	Litres	Liquid quarts	Litres
750	709. 750	800	757. 066	850	804. 383	900	851. 700	950	899. 016
1	710. 696	1	758. 013	1	805. 329	1	852. 646	1	899. 963
2	711. 642	2	758. 959	2	806. 276	2	853. 592	2	900. 909
3	712. 589	3	759. 905	3	807. 222	3	854. 539	3	901. 855
4	713. 535	4	760. 852	4	808. 168	4	855. 485	4	902. 802
5	714. 481	5	761. 798	5	809. 115	5	856. 431	5	903. 748
6	715. 428	6	762. 744	6	810. 061	6	857. 378	6	904. 694
7	716. 374	7	763. 691	7	811. 007	7	858. 324	7	905. 641
8	717. 320	8	764. 637	8	811. 954	8	859. 270	8	906. 587
9	718. 267	9	765. 583	9	812. 900	9	860. 217	9	907. 533
760	719. 213	810	766. 530	860	813. 846	910	861. 163	960	908. 480
1	720. 159	1	767. 476	1	814. 793	1	862. 109	1	909. 426
2	721. 106	2	768. 422	2	815. 739	2	863. 056	2	910. 372
3	722. 052	3	769. 369	3	816. 685	3	864. 002	3	911. 319
4	722. 998	4	770. 315	4	817. 632	4	864. 948	4	912. 265
5	723. 945	5	771. 261	5	818. 578	5	865. 895	5	913. 211
6	724. 891	6	772. 208	6	819. 524	6	866. 841	6	914. 158
7	725. 837	7	773. 154	7	820. 471	7	867. 787	7	915. 104
8	726. 784	8	774. 100	8	821. 417	8	868. 734	8	916. 050
9	727. 730	9	775. 047	9	822. 363	9	869. 680	9	916. 997
770	728. 676	820	775. 993	870	823. 310	920	870. 626	970	917. 943
1	729. 623	1	776. 939	1	824. 256	1	871. 573	1	918. 889
2	730. 569	2	777. 886	2	825. 202	2	872. 519	2	919. 836
3	731. 515	3	778. 832	3	826. 149	3	873. 465	3	920. 782
4	732. 462	4	779. 778	4	827. 095	4	874. 412	4	921. 728
5	733. 408	5	780. 725	5	828. 041	5	875. 358	5	922. 675
6	734. 354	6	781. 671	6	828. 988	6	876. 304	6	923. 621
7	735. 301	7	782. 617	7	829. 934	7	877. 251	7	924. 567
8	736. 247	8	783. 564	8	830. 880	8	878. 197	8	925. 514
9	737. 193	9	784. 510	9	831. 827	9	879. 143	9	926. 460
780	738. 140	830	785. 456	880	832. 773	930	880. 090	980	927. 406
1	739. 086	1	786. 403	1	833. 719	1	881. 036	1	928. 353
2	740. 032	2	787. 349	2	834. 666	2	881. 982	2	929. 299
3	740. 979	3	788. 295	3	835. 612	3	882. 929	3	930. 245
4	741. 925	4	789. 242	4	836. 558	4	883. 875	4	931. 192
5	742. 871	5	790. 188	5	837. 505	5	884. 821	5	932. 138
6	743. 818	6	791. 134	6	838. 451	6	885. 768	6	933. 084
7	744. 764	7	792. 081	7	839. 397	7	886. 714	7	934. 031
8	745. 710	8	793. 027	8	840. 344	8	887. 660	8	934. 977
9	746. 657	9	793. 973	9	841. 290	9	888. 607	9	935. 923
790	747. 603	840	794. 920	890	842. 236	940	889. 553	990	936. 870
1	748. 549	1	795. 866	1	843. 183	1	890. 499	1	937. 816
2	749. 496	2	796. 812	2	844. 129	2	891. 446	2	938. 762
3	750. 442	3	797. 759	3	845. 075	3	892. 392	3	939. 709
4	751. 388	4	798. 705	4	846. 022	4	893. 338	4	940. 655
5	752. 335	5	799. 651	5	846. 968	5	894. 285	5	941. 601
6	753. 281	6	800. 598	6	847. 914	6	895. 231	6	942. 548
7	754. 227	7	801. 544	7	848. 861	7	896. 177	7	943. 494
8	755. 174	8	802. 490	8	849. 807	8	897. 124	8	944. 440
9	756. 120	9	803. 437	9	850. 753	9	898. 070	9	945. 387

CAPACITY—LITRES TO GALLONS

[Reduction factor: 1 litre = 0.26417760 gallon]

Litres	Gallons	Litres	Gallons	Litres	Gallons	Litres	Gallons	Litres	Gallons
0		50	13.20888	100	26.4178	150	39.6266	200	52.8355
1	0.26418	1	13.47306	1	26.6819	1	39.8908	1	53.0997
2	0.52836	2	13.73724	2	26.9461	2	40.1550	2	53.3639
3	0.79253	3	14.00141	3	27.2103	3	40.4192	3	53.6281
4	1.05671	4	14.26559	4	27.4745	4	40.6834	4	53.8922
5	1.32089	5	14.52977	5	27.7386	5	40.9475	5	54.1564
6	1.58507	6	14.79395	6	28.0028	6	41.2117	6	54.4206
7	1.84924	7	15.05812	7	28.2670	7	41.4759	7	54.6848
8	2.11342	8	15.32230	8	28.5312	8	41.7401	8	54.9489
9	2.37760	9	15.58648	9	28.7954	9	42.0042	9	55.2131
10	2.64178	60	15.85066	110	29.0595	160	42.2684	210	55.4773
1	2.90595	1	16.11483	1	29.3237	1	42.5326	1	55.7415
2	3.17013	2	16.37901	2	29.5879	2	42.7968	2	56.0057
3	3.43431	3	16.64319	3	29.8521	3	43.0609	3	56.2698
4	3.69849	4	16.90737	4	30.1162	4	43.3251	4	56.5340
5	3.96266	5	17.17154	5	30.3804	5	43.5893	5	56.7982
6	4.22684	6	17.43572	6	30.6446	6	43.8535	6	57.0624
7	4.49102	7	17.69990	7	30.9088	7	44.1177	7	57.3265
8	4.75520	8	17.96408	8	31.1730	8	44.3818	8	57.5907
9	5.01937	9	18.22825	9	31.4371	9	44.6460	9	57.8549
20	5.28355	70	18.49243	120	31.7013	170	44.9102	220	58.1191
1	5.54773	1	18.75661	1	31.9655	1	45.1744	1	58.3832
2	5.81191	2	19.02079	2	32.2297	2	45.4385	2	58.6474
3	6.07608	3	19.28496	3	32.4938	3	45.7027	3	58.9116
4	6.34026	4	19.54914	4	32.7580	4	45.9669	4	59.1758
5	6.60444	5	19.81332	5	33.0222	5	46.2311	5	59.4400
6	6.86862	6	20.07750	6	33.2864	6	46.4953	6	59.7041
7	7.13280	7	20.34168	7	33.5506	7	46.7594	7	59.9683
8	7.39697	8	20.60585	8	33.8147	8	47.0236	8	60.2325
9	7.66115	9	20.87003	9	34.0789	9	47.2878	9	60.4967
30	7.92533	80	21.13421	130	34.3431	180	47.5520	230	60.7608
1	8.18951	1	21.39839	1	34.6073	1	47.8161	1	61.0250
2	8.45368	2	21.66256	2	34.8714	2	48.0803	2	61.2892
3	8.71786	3	21.92674	3	35.1356	3	48.3445	3	61.5534
4	8.98204	4	22.19092	4	35.3998	4	48.6087	4	61.8176
5	9.24622	5	22.45510	5	35.6640	5	48.8729	5	62.0817
6	9.51039	6	22.71927	6	35.9282	6	49.1370	6	62.3459
7	9.77457	7	22.98345	7	36.1923	7	49.4012	7	62.6101
8	10.03875	8	23.24763	8	36.4565	8	49.6654	8	62.8743
9	10.30293	9	23.51181	9	36.7207	9	49.9296	9	63.1384
40	10.56710	90	23.77598	140	36.9849	190	50.1937	240	63.4026
1	10.83128	1	24.04016	1	37.2490	1	50.4579	1	63.6668
2	11.09546	2	24.30434	2	37.5132	2	50.7221	2	63.9310
3	11.35964	3	24.56852	3	37.7774	3	50.9863	3	64.1952
4	11.62381	4	24.83269	4	38.0416	4	51.2505	4	64.4593
5	11.88799	5	25.09687	5	38.3058	5	51.5146	5	64.7235
6	12.15217	6	25.36105	6	38.5699	6	51.7788	6	64.9877
7	12.41635	7	25.62523	7	38.8341	7	52.0430	7	65.2519
8	12.68052	8	25.88940	8	39.0983	8	52.3072	8	65.5160
9	12.94470	9	26.15358	9	39.3625	9	52.5713	9	65.7802

CAPACITY—LITRES TO GALLONS

[Reduction factor: 1 litre = 0.26417760 gallon]

Litres	Gallons	Litres	Gallons	Litres	Gallons	Litres	Gallons	Litres	Gallons
250	66.0444	300	79.2533	350	92.4622	400	105.6710	450	118.8799
1	66.3086	1	79.5175	1	92.7263	1	105.9352	1	119.1441
2	66.5728	2	79.7816	2	92.9905	2	106.1994	2	119.4083
3	66.8369	3	80.0458	3	93.2547	3	106.4636	3	119.6725
4	67.1011	4	80.3100	4	93.5189	4	106.7278	4	119.9366
5	67.3653	5	80.5742	5	93.7830	5	106.9919	5	120.2008
6	67.6295	6	80.8383	6	94.0472	6	107.2561	6	120.4650
7	67.8936	7	81.1025	7	94.3114	7	107.5203	7	120.7292
8	68.1578	8	81.3667	8	94.5756	8	107.7845	8	120.9933
9	68.4220	9	81.6309	9	94.8398	9	108.0486	9	121.2575
260	68.6862	310	81.8951	360	95.1039	410	108.3128	460	121.5217
1	68.9504	1	82.1592	1	95.3681	1	108.5770	1	121.7859
2	69.2145	2	82.4234	2	95.6323	2	108.8412	2	122.0501
3	69.4787	3	82.6876	3	95.8965	3	109.1053	3	122.3142
4	69.7429	4	82.9518	4	96.1606	4	109.3695	4	122.5784
5	70.0071	5	83.2159	5	96.4248	5	109.6337	5	122.8426
6	70.2712	6	83.4801	6	96.6890	6	109.8979	6	123.1068
7	70.5354	7	83.7443	7	96.9532	7	110.1621	7	123.3709
8	70.7996	8	84.0085	8	97.2174	8	110.4262	8	123.6351
9	71.0638	9	84.2727	9	97.4815	9	110.6904	9	123.8993
270	71.3280	320	84.5368	370	97.7457	420	110.9546	470	124.1635
1	71.5921	1	84.8010	1	98.0099	1	111.2188	1	124.4276
2	71.8563	2	85.0652	2	98.2741	2	111.4829	2	124.6918
3	72.1205	3	85.3294	3	98.5382	3	111.7471	3	124.9560
4	72.3847	4	85.5935	4	98.8024	4	112.0113	4	125.2202
5	72.6488	5	85.8577	5	99.0666	5	112.2755	5	125.4844
6	72.9130	6	86.1219	6	99.3308	6	112.5397	6	125.7485
7	73.1772	7	86.3861	7	99.5950	7	112.8038	7	126.0127
8	73.4414	8	86.6503	8	99.8591	8	113.0680	8	126.2769
9	73.7056	9	86.9144	9	100.1233	9	113.3322	9	126.5411
280	73.9697	330	87.1786	380	100.3875	430	113.5964	480	126.8052
1	74.2339	1	87.4428	1	100.6517	1	113.8605	1	127.0694
2	74.4981	2	87.7070	2	100.9158	2	114.1247	2	127.3336
3	74.7623	3	87.9711	3	101.1800	3	114.3889	3	127.5978
4	75.0264	4	88.2353	4	101.4442	4	114.6531	4	127.8620
5	75.2906	5	88.4995	5	101.7084	5	114.9173	5	128.1261
6	75.5548	6	88.7637	6	101.9726	6	115.1814	6	128.3903
7	75.8190	7	89.0279	7	102.2367	7	115.4456	7	128.6545
8	76.0831	8	89.2920	8	102.5009	8	115.7098	8	128.9187
9	76.3473	9	89.5562	9	102.7651	9	115.9740	9	129.1828
290	76.6115	340	89.8204	390	103.0293	440	116.2381	490	129.4470
1	76.8757	1	90.0846	1	103.2934	1	116.5023	1	129.7112
2	77.1399	2	90.3487	2	103.5576	2	116.7665	2	129.9754
3	77.4040	3	90.6129	3	103.8218	3	117.0307	3	130.2396
4	77.6682	4	90.8771	4	104.0860	4	117.2949	4	130.5037
5	77.9324	5	91.1413	5	104.3502	5	117.5590	5	130.7679
6	78.1966	6	91.4054	6	104.6143	6	117.8232	6	131.0321
7	78.4607	7	91.6696	7	104.8785	7	118.0874	7	131.2963
8	78.7249	8	91.9338	8	105.1427	8	118.3516	8	131.5604
9	78.9891	9	92.1980	9	105.4069	9	118.6157	9	131.8246

CAPACITY—LITRES TO GALLONS

[Reduction factor: 1 litre = 0.26417760 gallon]

Litres	Gallons	Litres	Gallons	Litres	Gallons	Litres	Gallons	Litres	Gallons
500	132.0888	550	145.2977	600	158.5066	650	171.7154	700	184.9243
1	132.3530	1	145.5619	1	158.7707	1	171.9796	1	185.1885
2	132.6172	2	145.8260	2	159.0349	2	172.2438	2	185.4527
3	132.8813	3	146.0902	3	159.2991	3	172.5080	3	185.7169
4	133.1455	4	146.3544	4	159.5633	4	172.7722	4	185.9810
5	133.4097	5	146.6186	5	159.8274	5	173.0363	5	186.2452
6	133.6739	6	146.8827	6	160.0916	6	173.3005	6	186.5094
7	133.9380	7	147.1469	7	160.3558	7	173.5647	7	186.7736
8	134.2022	8	147.4111	8	160.6200	8	173.8289	8	187.0377
9	134.4664	9	147.6753	9	160.8842	9	174.0930	9	187.3019
510	134.7306	560	147.9395	610	161.1483	660	174.3572	710	187.5661
1	134.9948	1	148.2036	1	161.4125	1	174.6214	1	187.8303
2	135.2589	2	148.4678	2	161.6767	2	174.8856	2	188.0945
3	135.5231	3	148.7320	3	161.9409	3	175.1497	3	188.3586
4	135.7873	4	148.9962	4	162.2050	4	175.4139	4	188.6228
5	136.0515	5	149.2603	5	162.4692	5	175.6781	5	188.8870
6	136.3156	6	149.5245	6	162.7334	6	175.9423	6	189.1512
7	136.5798	7	149.7887	7	162.9976	7	176.2065	7	189.4153
8	136.8440	8	150.0529	8	163.2618	8	176.4706	8	189.6795
9	137.1082	9	150.3171	9	163.5259	9	176.7348	9	189.9437
520	137.3724	570	150.5812	620	163.7901	670	176.9990	720	190.2079
1	137.6365	1	150.8454	1	164.0543	1	177.2632	1	190.4720
2	137.9007	2	151.1096	2	164.3185	2	177.5273	2	190.7362
3	138.1649	3	151.3738	3	164.5826	3	177.7915	3	191.0004
4	138.4291	4	151.6379	4	164.8468	4	178.0557	4	191.2646
5	138.6932	5	151.9021	5	165.1110	5	178.3199	5	191.5288
6	138.9574	6	152.1663	6	165.3752	6	178.5841	6	191.7929
7	139.2216	7	152.4305	7	165.6394	7	178.8482	7	192.0571
8	139.4858	8	152.6947	8	165.9035	8	179.1124	8	192.3213
9	139.7500	9	152.9588	9	166.1677	9	179.3766	9	192.5855
530	140.0141	580	153.2230	630	166.4319	680	179.6408	730	192.8496
1	140.2783	1	153.4872	1	166.6961	1	179.9049	1	193.1138
2	140.5425	2	153.7514	2	166.9602	2	180.1691	2	193.3780
3	140.8067	3	154.0155	3	167.2244	3	180.4333	3	193.6422
4	141.0708	4	154.2797	4	167.4886	4	180.6975	4	193.9064
5	141.3350	5	154.5439	5	167.7528	5	180.9617	5	194.1705
6	141.5992	6	154.8081	6	168.0170	6	181.2258	6	194.4347
7	141.8634	7	155.0723	7	168.2811	7	181.4900	7	194.6989
8	142.1275	8	155.3364	8	168.5453	8	181.7542	8	194.9631
9	142.3917	9	155.6006	9	168.8095	9	182.0184	9	195.2272
540	142.6559	590	155.8648	640	169.0737	690	182.2825	740	195.4914
1	142.9201	1	156.1290	1	169.3378	1	182.5467	1	195.7556
2	143.1843	2	156.3931	2	169.6020	2	182.8109	2	196.0198
3	143.4484	3	156.6573	3	169.8662	3	183.0751	3	196.2840
4	143.7126	4	156.9215	4	170.1304	4	183.3393	4	196.5481
5	143.9768	5	157.1857	5	170.3946	5	183.6034	5	196.8123
6	144.2410	6	157.4498	6	170.6587	6	183.8676	6	197.0765
7	144.5051	7	157.7140	7	170.9229	7	184.1318	7	197.3407
8	144.7693	8	157.9782	8	171.1871	8	184.3960	8	197.6048
9	145.0335	9	158.2424	9	171.4513	9	184.6601	9	197.8690

CAPACITY—LITRES TO GALLONS

[Reduction factor: 1 litre = 0.26417760 gallon]

Litres	Gallons	Litres	Gallons	Litres	Gallons	Litres	Gallons	Litres	Gallons
750	198.1332	800	211.3421	850	224.5510	900	237.7598	950	250.9687
1	198.3974	1	211.6063	1	224.8151	1	238.0240	1	251.2329
2	198.6616	2	211.8704	2	225.0793	2	238.2882	2	251.4971
3	198.9257	3	212.1346	3	225.3435	3	238.5524	3	251.7613
4	199.1899	4	212.3988	4	225.6077	4	238.8166	4	252.0254
5	199.4541	5	212.6630	5	225.8718	5	239.0807	5	252.2896
6	199.7183	6	212.9271	6	226.1360	6	239.3449	6	252.5538
7	199.9824	7	213.1913	7	226.4002	7	239.6091	7	252.8180
8	200.2466	8	213.4555	8	226.6644	8	239.8733	8	253.0821
9	200.5108	9	213.7197	9	226.9286	9	240.1374	9	253.3463
760	200.7750	810	213.9839	860	227.1927	910	240.4016	960	253.6105
1	201.0392	1	214.2480	1	227.4569	1	240.6658	1	253.8747
2	201.3033	2	214.5122	2	227.7211	2	240.9300	2	254.1389
3	201.5675	3	214.7764	3	227.9853	3	241.1941	3	254.4030
4	201.8317	4	215.0406	4	228.2494	4	241.4583	4	254.6672
5	202.0959	5	215.3047	5	228.5136	5	241.7225	5	254.9314
6	202.3600	6	215.5689	6	228.7778	6	241.9867	6	255.1956
7	202.6242	7	215.8331	7	229.0420	7	242.2509	7	255.4597
8	202.8884	8	216.0973	8	229.3062	8	242.5150	8	255.7239
9	203.1526	9	216.3615	9	229.5703	9	242.7792	9	255.9881
770	203.4168	820	216.6256	870	229.8345	920	243.0434	970	256.2523
1	203.6809	1	216.8898	1	230.0987	1	243.3076	1	256.5164
2	203.9451	2	217.1540	2	230.3629	2	243.5717	2	256.7806
3	204.2093	3	217.4182	3	230.6270	3	243.8359	3	257.0448
4	204.4735	4	217.6823	4	230.8912	4	244.1001	4	257.3090
5	204.7376	5	217.9465	5	231.1554	5	244.3643	5	257.5732
6	205.0018	6	218.2107	6	231.4196	6	244.6285	6	257.8373
7	205.2660	7	218.4749	7	231.6838	7	244.8926	7	258.1015
8	205.5302	8	218.7391	8	231.9479	8	245.1568	8	258.3657
9	205.7944	9	219.0032	9	232.2121	9	245.4210	9	258.6299
780	206.0585	830	219.2674	880	232.4763	930	245.6852	980	258.8940
1	206.3227	1	219.5316	1	232.7405	1	245.9493	1	259.1582
2	206.5869	2	219.7958	2	233.0046	2	246.2135	2	259.4224
3	206.8511	3	220.0599	3	233.2688	3	246.4777	3	259.6866
4	207.1152	4	220.3241	4	233.5330	4	246.7419	4	259.9508
5	207.3794	5	220.5883	5	233.7972	5	247.0061	5	260.2149
6	207.6436	6	220.8525	6	234.0614	6	247.2702	6	260.4791
7	207.9078	7	221.1167	7	234.3255	7	247.5344	7	260.7433
8	208.1719	8	221.3808	8	234.5897	8	247.7986	8	261.0075
9	208.4361	9	221.6450	9	234.8539	9	248.0628	9	261.2716
790	208.7003	840	221.9092	890	235.1181	940	248.3269	990	261.5358
1	208.9645	1	222.1734	1	235.3822	1	248.5911	1	261.8000
2	209.2287	2	222.4375	2	235.6464	2	248.8553	2	262.0642
3	209.4928	3	222.7017	3	235.9106	3	249.1195	3	262.3284
4	209.7570	4	222.9659	4	236.1748	4	249.3837	4	262.5925
5	210.0212	5	223.2301	5	236.4390	5	249.6478	5	262.8567
6	210.2854	6	223.4942	6	236.7031	6	249.9120	6	263.1209
7	210.5495	7	223.7584	7	236.9673	7	250.1762	7	263.3851
8	210.8137	8	224.0226	8	237.2315	8	250.4404	8	263.6492
9	211.0779	9	224.2868	9	237.4957	9	250.7045	9	263.9134

CAPACITY—GALLONS TO LITRES

[Reduction factor: 1 gallon = 3.7853323 litres]

Gallons	Litres	Gallons	Litres	Gallons	Litres	Gallons	Litres	Gallons	Litres
0		50	189.2666	100	378.533	150	567.800	200	757.066
1	3.7853	1	193.0519	1	382.319	1	571.585	1	760.852
2	7.5707	2	196.8373	2	386.104	2	575.371	2	764.637
3	11.3560	3	200.6226	3	389.889	3	579.156	3	768.422
4	15.1413	4	204.4079	4	393.675	4	582.941	4	772.208
5	18.9267	5	208.1933	5	397.460	5	586.727	5	775.993
6	22.7120	6	211.9786	6	401.245	6	590.512	6	779.778
7	26.4973	7	215.7639	7	405.031	7	594.297	7	783.564
8	30.2827	8	219.5493	8	408.816	8	598.083	8	787.249
9	34.0680	9	223.3346	9	412.601	9	601.868	9	791.134
10	37.8533	60	227.1199	110	416.387	160	605.653	210	794.920
1	41.6387	1	230.9053	1	420.172	1	609.438	1	798.705
2	45.4240	2	234.6906	2	423.957	2	613.224	2	802.490
3	49.2093	3	238.4759	3	427.743	3	617.009	3	806.276
4	52.9947	4	242.2613	4	431.528	4	620.794	4	810.061
5	56.7800	5	246.0466	5	435.313	5	624.580	5	813.846
6	60.5653	6	249.8319	6	439.099	6	628.365	6	817.632
7	64.3506	7	253.6173	7	442.884	7	632.150	7	821.417
8	68.1360	8	257.4026	8	446.669	8	635.936	8	825.202
9	71.9213	9	261.1879	9	450.455	9	639.721	9	828.988
20	75.7066	70	264.9733	120	454.240	170	643.506	220	832.773
1	79.4920	1	268.7586	1	458.025	1	647.292	1	836.558
2	83.2773	2	272.5439	2	461.811	2	651.077	2	840.344
3	87.0626	3	276.3293	3	465.596	3	654.862	3	844.129
4	90.8480	4	280.1146	4	469.381	4	658.648	4	847.914
5	94.6333	5	283.8999	5	473.167	5	662.433	5	851.700
6	98.4186	6	287.6853	6	476.952	6	666.218	6	855.485
7	102.2040	7	291.4706	7	480.737	7	670.004	7	859.270
8	105.9893	8	295.2559	8	484.523	8	673.789	8	863.056
9	109.7746	9	299.0413	9	488.308	9	677.574	9	866.841
30	113.5600	80	302.8266	130	492.093	180	681.360	230	870.626
1	117.3453	1	306.6119	1	495.879	1	685.145	1	874.412
2	121.1306	2	310.3972	2	499.664	2	688.930	2	878.197
3	124.9160	3	314.1826	3	503.449	3	692.716	3	881.982
4	128.7013	4	317.9679	4	507.235	4	696.501	4	885.768
5	132.4866	5	321.7532	5	511.020	5	700.286	5	889.553
6	136.2720	6	325.5386	6	514.805	6	704.072	6	893.338
7	140.0573	7	329.3239	7	518.591	7	707.857	7	897.124
8	143.8426	8	333.1092	8	522.376	8	711.642	8	900.909
9	147.6280	9	336.8946	9	526.161	9	715.428	9	904.694
40	151.4133	90	340.6799	140	529.947	190	719.213	240	908.480
1	155.1986	1	344.4652	1	533.732	1	722.998	1	912.265
2	158.9840	2	348.2506	2	537.517	2	726.784	2	916.050
3	162.7693	3	352.0359	3	541.303	3	730.569	3	919.836
4	166.5546	4	355.8212	4	545.088	4	734.354	4	923.621
5	170.3400	5	359.6066	5	548.873	5	738.140	5	927.406
6	174.1253	6	363.3919	6	552.659	6	741.925	6	931.192
7	177.9106	7	367.1772	7	556.444	7	745.710	7	934.977
8	181.6960	8	370.9626	8	560.229	8	749.496	8	938.762
9	185.4813	9	374.7479	9	564.015	9	753.281	9	942.548

CAPACITY—GALLONS TO LITRES

[Reduction factor: 1 gallon = 3.7853323 litres]

Gallons	Litres	Gallons	Litres	Gallons	Litres	Gallons	Litres	Gallons	Litres
250	946. 333	**300**	1,135. 600	**350**	1,324. 866	**400**	1,514. 133	**450**	1,703. 400
1	950. 118	1	1,139. 385	1	1,328. 652	1	1,517. 918	1	1,707. 185
2	953. 904	2	1,143. 170	2	1,332. 437	2	1,521. 704	2	1,710. 970
3	957. 689	3	1,146. 956	3	1,336. 222	3	1,525. 489	3	1,714. 756
4	961. 474	4	1,150. 741	4	1,340. 008	4	1,529. 274	4	1,718. 541
5	965. 260	5	1,154. 526	5	1,343. 793	5	1,533. 060	5	1,722. 326
6	969. 045	6	1,158. 312	6	1,347. 578	6	1,536. 845	6	1,776. 112
7	972. 830	7	1,162. 097	7	1,351. 364	7	1,540. 630	7	1,729. 897
8	976. 616	8	1,165. 882	8	1,355. 149	8	1,544. 416	8	1,733. 682
9	980. 401	9	1,169. 668	9	1,358. 934	9	1,548. 201	9	1,737. 468
260	984. 186	**310**	1,173. 453	**360**	1,362. 720	**410**	1,551. 986	**460**	1,741. 253
1	987. 972	1	1,177. 238	1	1,366. 525	1	1,555. 772	1	1,745. 038
2	991. 757	2	1,181. 024	2	1,370. 290	2	1,559. 557	2	1,748. 824
3	995. 542	3	1,184. 809	3	1,374. 076	3	1,563. 342	3	1,752. 609
4	999. 328	4	1,188. 594	4	1,377. 861	4	1,567. 128	4	1,756. 394
5	1,003. 113	5	1,192. 380	5	1,381. 646	5	1,570. 913	5	1,760. 180
6	1,006. 898	6	1,196. 165	6	1,385. 432	6	1,574. 698	6	1,763. 965
7	1,010. 684	7	1,199. 950	7	1,389. 217	7	1,578. 484	7	1,767. 750
8	1,014. 469	8	1,203. 736	8	1,393. 002	8	1,582. 269	8	1,771. 536
9	1,018. 254	9	1,207. 521	9	1,396. 788	9	1,586. 054	9	1,775. 321
270	1,022. 040	**320**	1,211. 306	**370**	1,400. 573	**420**	1,589. 840	**470**	1,779. 106
1	1,025. 825	1	1,215. 092	1	1,404. 358	1	1,593. 625	1	1,782. 892
2	1,029. 610	2	1,218. 877	2	1,408. 144	2	1,597. 410	2	1,786. 677
3	1,033. 396	3	1,222. 662	3	1,411. 929	3	1,601. 196	3	1,790. 462
4	1,037. 181	4	1,226. 448	4	1,415. 714	4	1,604. 981	4	1,794. 248
5	1,040. 966	5	1,230. 233	5	1,419. 500	5	1,608. 766	5	1,798. 033
6	1,044. 752	6	1,234. 018	6	1,423. 285	6	1,612. 552	6	1,801. 818
7	1,048. 537	7	1,237. 804	7	1,427. 070	7	1,616. 337	7	1,805. 604
8	1,052. 322	8	1,241. 589	8	1,430. 856	8	1,620. 122	8	1,809. 389
9	1,056. 108	9	1,245. 374	9	1,434. 641	9	1,623. 908	9	1,813. 174
280	1,059. 893	**330**	1,249. 160	**380**	1,438. 426	**430**	1,627. 693	**480**	1,816. 960
1	1,063. 678	1	1,252. 945	1	1,442. 212	1	1,631. 478	1	1,820. 745
2	1,067. 464	2	1,256. 730	2	1,445. 997	2	1,635. 264	2	1,824. 530
3	1,071. 249	3	1,260. 516	3	1,449. 782	3	1,639. 049	3	1,828. 316
4	1,075. 034	4	1,264. 301	4	1,453. 568	4	1,642. 834	4	1,832. 101
5	1,078. 820	5	1,268. 086	5	1,457. 353	5	1,646. 620	5	1,835. 886
6	1,082. 605	6	1,271. 872	6	1,461. 138	6	1,650. 405	6	1,839. 671
7	1,086. 390	7	1,275. 657	7	1,464. 924	7	1,654. 190	7	1,843. 457
8	1,090. 176	8	1,279. 442	8	1,468. 709	8	1,657. 976	8	1,847. 242
9	1,093. 961	9	1,283. 228	9	1,472. 494	9	1,661. 761	9	1,851. 027
290	1,097. 746	**340**	1,287. 013	**390**	1,476. 280	**440**	1,665. 546	**490**	1,854. 813
1	1,101. 532	1	1,290. 798	1	1,480. 065	1	1,669. 332	1	1,858. 598
2	1,105. 317	2	1,294. 584	2	1,483. 850	2	1,673. 117	2	1,862. 383
3	1,109. 102	3	1,298. 369	3	1,487. 636	3	1,676. 902	3	1,866. 169
4	1,112. 888	4	1,302. 154	4	1,491. 421	4	1,680. 688	4	1,869. 954
5	1,116. 673	5	1,305. 940	5	1,495. 206	5	1,684. 473	5	1,873. 739
6	1,120. 458	6	1,309. 725	6	1,498. 992	6	1,688. 258	6	1,877. 525
7	1,124. 244	7	1,313. 510	7	1,502. 777	7	1,692. 044	7	1,881. 310
8	1,128. 029	8	1,317. 296	8	1,506. 562	8	1,695. 829	8	1,885. 095
9	1,131. 814	9	1,321. 081	9	1,510. 348	9	1,699. 614	9	1,888. 881

CAPACITY—GALLONS TO LITRES

[Reduction factor: 1 gallon = 3.7853323 litres]

Gallons	Litres	Gallons	Litres	Gallons	Litres	Gallons	Litres	Gallons	Litres
500	1,892.666	550	2,081.933	600	2,271.199	650	2,460.466	700	2,649.733
1	1,896.451	1	2,085.718	1	2,274.985	1	2,464.251	1	2,653.518
2	1,900.237	2	2,089.503	2	2,278.770	2	2,468.037	2	2,657.303
3	1,904.022	3	2,093.289	3	2,282.555	3	2,471.822	3	2,661.089
4	1,907.807	4	2,097.074	4	2,286.341	4	2,475.607	4	2,664.874
5	1,911.593	5	2,100.859	5	2,290.126	5	2,479.393	5	2,668.659
6	1,915.378	6	2,104.645	6	2,293.911	6	2,483.178	6	2,672.445
7	1,919.163	7	2,108.430	7	2,297.697	7	2,486.963	7	2,676.230
8	1,922.949	8	2,112.215	8	2,301.482	8	2,490.749	8	2,680.015
9	1,926.734	9	2,116.001	9	2,305.267	9	2,494.534	9	2,683.801
510	1,930.519	560	2,119.786	610	2,309.053	660	2,498.319	710	2,687.586
1	1,934.305	1	2,123.571	1	2,312.838	1	2,502.105	1	2,691.371
2	1,938.090	2	2,127.357	2	2,316.623	2	2,505.890	2	2,695.157
3	1,941.875	3	2,131.142	3	2,320.409	3	2,509.675	3	2,698.942
4	1,945.661	4	2,134.927	4	2,324.194	4	2,513.461	4	2,702.727
5	1,949.446	5	2,138.713	5	2,327.979	5	2,517.246	5	2,706.513
6	1,953.231	6	2,142.498	6	2,331.765	6	2,521.031	6	2,710.298
7	1,957.017	7	2,146.283	7	2,335.550	7	2,524.817	7	2,714.083
8	1,960.802	8	2,150.069	8	2,339.335	8	2,528.602	8	2,717.869
9	1,964.587	9	2,153.854	9	2,343.121	9	2,532.387	9	2,721.654
520	1,968.373	570	2,157.639	620	2,346.906	670	2,536.173	720	2,725.439
1	1,972.158	1	2,161.425	1	2,350.691	1	2,539.958	1	2,729.225
2	1,975.943	2	2,165.210	2	2,354.477	2	2,543.743	2	2,733.010
3	1,979.729	3	2,168.995	3	2,358.262	3	2,547.529	3	2,736.795
4	1,983.514	4	2,172.781	4	2,362.047	4	2,551.314	4	2,740.581
5	1,987.299	5	2,176.566	5	2,365.833	5	2,555.099	5	2,744.366
6	1,991.085	6	2,180.351	6	2,369.618	6	2,558.885	6	2,748.151
7	1,994.870	7	2,184.137	7	2,373.403	7	2,562.670	7	2,751.937
8	1,998.655	8	2,187.922	8	2,377.189	8	2,566.455	8	2,755.722
9	2,002.441	9	2,191.707	9	2,380.974	9	2,570.241	9	2,759.507
530	2,006.226	580	2,195.493	630	2,384.759	680	2,574.026	730	2,763.293
1	2,010.011	1	2,199.278	1	2,388.545	1	2,577.811	1	2,767.078
2	2,013.797	2	2,203.063	2	2,392.330	2	2,581.597	2	2,770.863
3	2,017.582	3	2,206.849	3	2,396.115	3	2,585.382	3	2,774.649
4	2,021.367	4	2,210.634	4	2,399.901	4	2,589.167	4	2,778.434
5	2,025.153	5	2,214.419	5	2,403.686	5	2,592.953	5	2,782.219
6	2,028.938	6	2,218.205	6	2,407.471	6	2,596.738	6	2,786.005
7	2,032.723	7	2,221.990	7	2,411.257	7	2,600.523	7	2,789.790
8	2,036.509	8	2,225.775	8	2,415.042	8	2,604.309	8	2,793.575
9	2,040.294	9	2,229.561	9	2,418.827	9	2,608.094	9	2,797.361
540	2,044.079	590	2,233.346	640	2,422.613	690	2,611.879	740	2,801.146
1	2,047.865	1	2,237.131	1	2,426.398	1	2,615.665	1	2,804.931
2	2,051.650	2	2,240.917	2	2,430.183	2	2,619.450	2	2,808.717
3	2,055.435	3	2,244.702	3	2,433.969	3	2,623.235	3	2,812.502
4	2,059.221	4	2,248.487	4	2,437.754	4	2,627.021	4	2,816.287
5	2,063.006	5	2,252.273	5	2,441.539	5	2,630.806	5	2,820.073
6	2,066.791	6	2,256.058	6	2,445.325	6	2,634.591	6	2,823.858
7	2,070.577	7	2,259.843	7	2,449.110	7	2,638.377	7	2,827.643
8	2,074.362	8	2,263.629	8	2,452.895	8	2,642.162	8	2,831.429
9	2,078.147	9	2,267.414	9	2,456.681	9	2,645.947	9	2,835.214

CAPACITY—GALLONS TO LITRES

[Reduction factor: 1 gallon = 3.7853323 litres]

Gallons	Litres	Gallons	Litres	Gallons	Litres	Gallons	Litres	Gallons	Litres
750	2,838.999	800	3,028.266	850	3,217.532	900	3,406.799	950	3,596.066
1	2,842.785	1	3,032.051	1	3,221.318	1	3,410.584	1	3,599.851
2	2,846.570	2	3,035.837	2	3,225.103	2	3,414.370	2	3,603.636
3	2,850.355	3	3,039.622	3	3,228.888	3	3,418.155	3	3,607.422
4	2,854.141	4	3,043.407	4	3,232.674	4	3,421.940	4	3,611.207
5	2,857.926	5	3,047.193	5	3,236.459	5	3,425.726	5	3,614.992
6	2,861.711	6	3,050.978	6	3,240.244	6	3,429.511	6	3,618.778
7	2,865.497	7	3,054.763	7	3,244.030	7	3,433.296	7	3,622.563
8	2,869.282	8	3,058.548	8	3,247.815	8	3,437.082	8	3,626.348
9	2,873.067	9	3,062.334	9	3,251.600	9	3,440.867	9	3,630.134
760	2,876.853	810	3,066.119	860	3,255.386	910	3,444.652	960	3,633.919
1	2,880.638	1	3,069.904	1	3,259.171	1	3,448.438	1	3,637.704
2	2,884.423	2	3,073.690	2	3,262.956	2	3,452.223	2	3,641.490
3	2,888.209	3	3,077.475	3	3,266.742	3	3,456.008	3	3,645.275
4	2,891.994	4	3,081.260	4	3,270.527	4	3,459.794	4	3,649.060
5	2,895.779	5	3,085.046	5	3,274.312	5	3,463.579	5	3,652.846
6	2,899.565	6	3,088.831	6	3,278.098	6	3,467.364	6	3,656.631
7	2,903.350	7	3,092.616	7	3,281.883	7	3,471.150	7	3,660.416
8	2,907.135	8	3,096.402	8	3,285.668	8	3,474.935	8	3,664.202
9	2,910.921	9	3,100.187	9	3,289.454	9	3,478.720	9	3,667.987
770	2,914.706	820	3,103.972	870	3,293.239	920	3,482.506	970	3,671.772
1	2,918.491	1	3,107.758	1	3,297.024	1	3,486.291	1	3,675.558
2	2,922.277	2	3,111.543	2	3,300.810	2	3,490.076	2	3,679.343
3	2,926.062	3	3,115.328	3	3,304.595	3	3,493.862	3	3,683.128
4	2,929.847	4	3,119.114	4	3,308.380	4	3,497.647	4	3,686.914
5	2,933.633	5	3,122.899	5	3,312.166	5	3,501.432	5	3,690.699
6	2,937.418	6	3,126.684	6	3,315.951	6	3,505.218	6	3,694.484
7	2,941.203	7	3,130.470	7	3,319.736	7	3,509.003	7	3,698.270
8	2,944.989	8	3,134.255	8	3,323.522	8	3,512.788	8	3,702.055
9	2,948.774	9	3,138.040	9	3,327.307	9	3,516.574	9	3,705.840
780	2,952.559	830	3,141.826	880	3,331.092	930	3,520.359	980	3,709.626
1	2,956.345	1	3,145.611	1	3,334.878	1	3,524.144	1	3,713.411
2	2,960.130	2	3,149.396	2	3,338.663	2	3,527.930	2	3,717.196
3	2,963.915	3	3,153.182	3	3,342.448	3	3,531.715	3	3,720.982
4	2,967.701	4	3,156.967	4	3,346.234	4	3,535.500	4	3,724.767
5	2,971.486	5	3,160.752	5	3,350.019	5	3,539.286	5	3,728.552
6	2,975.271	6	3,164.538	6	3,353.804	6	3,543.071	6	3,732.338
7	2,979.057	7	3,168.323	7	3,357.590	7	3,546.856	7	3,736.123
8	2,982.842	8	3,172.108	8	3,361.375	8	3,550.642	8	3,739.908
9	2,986.627	9	3,175.894	9	3,365.160	9	3,554.427	9	3,743.694
790	2,990.413	840	3,179.679	890	3,368.946	940	3,558.212	990	3,747.479
1	2,994.198	1	3,183.464	1	3,372.731	1	3,561.998	1	3,751.264
2	2,997.983	2	3,187.250	2	3,376.516	2	3,565.783	2	3,755.050
3	3,001.769	3	3,191.035	3	3,380.302	3	3,569.568	3	3,758.835
4	3,005.554	4	3,194.820	4	3,384.087	4	3,573.354	4	3,762.620
5	3,009.339	5	3,198.606	5	3,387.872	5	3,577.139	5	3,766.406
6	3,013.125	6	3,202.391	6	3,391.658	6	3,580.924	6	3,770.191
7	3,016.910	7	3,206.176	7	3,395.443	7	3,584.710	7	3,773.976
8	3,020.695	8	3,209.962	8	3,399.228	8	3,588.495	8	3,777.762
9	3,024.481	9	3,213.747	9	3,403.014	9	3,592.280	9	3,781.547

CAPACITY—HECTOLITRES TO BUSHELS

[Reduction factor: 1 hectolitre = 2.8378189 bushels]

Hecto-litres	Bush-els	Hecto-litres	Bush-els	Hecto-litres	Bush-els	Hecto-litres	Bush-els	Hecto-litres	Bush-els
0		50	141.8909	100	283.782	150	425.673	200	567.564
1	2.8378	1	144.7288	1	286.620	1	428.511	1	570.402
2	5.6756	2	147.5666	2	289.458	2	431.348	2	573.239
3	8.5135	3	150.4044	3	292.295	3	434.186	3	576.077
4	11.3513	4	153.2422	4	295.133	4	437.024	4	578.915
5	14.1891	5	156.0800	5	297.971	5	439.862	5	581.753
6	17.0269	6	158.9179	6	300.809	6	442.700	6	584.591
7	19.8647	7	161.7557	7	303.647	7	445.538	7	587.429
8	22.7026	8	164.5935	8	306.484	8	448.375	8	590.266
9	25.5404	9	167.4313	9	309.322	9	451.213	9	593.104
10	28.3782	60	170.2691	110	312.160	160	454.051	210	595.942
1	31.2160	1	173.1070	1	314.998	1	456.889	1	598.780
2	34.0538	2	175.9448	2	317.836	2	459.727	2	601.618
3	36.8916	3	178.7826	3	320.674	3	462.564	3	604.455
4	39.7295	4	181.6204	4	323.511	4	465.402	4	607.293
5	42.5673	5	184.4582	5	326.349	5	468.240	5	610.131
6	45.4051	6	187.2960	6	329.187	6	471.078	6	612.969
7	48.2429	7	190.1339	7	332.025	7	473.916	7	615.807
8	51.0807	8	192.9717	8	334.863	8	476.754	8	618.645
9	53.9186	9	195.8095	9	337.700	9	479.591	9	621.482
20	56.7564	70	198.6473	120	340.538	170	482.429	220	624.320
1	59.5942	1	201.4851	1	343.376	1	485.267	1	627.158
2	62.4320	2	204.3230	2	346.214	2	488.105	2	629.996
3	65.2698	3	207.1608	3	349.052	3	490.943	3	632.834
4	68.1077	4	209.9986	4	351.890	4	493.780	4	635.671
5	70.9455	5	212.8364	5	354.727	5	496.618	5	638.509
6	73.7833	6	215.6742	6	357.565	6	499.456	6	641.347
7	76.6211	7	218.5121	7	360.403	7	502.294	7	644.185
8	79.4589	8	221.3499	8	363.241	8	505.132	8	647.023
9	82.2967	9	224.1877	9	366.079	9	507.970	9	649.861
30	85.1346	80	227.0255	130	368.916	180	510.807	230	652.698
1	87.9724	1	229.8633	1	371.754	1	513.645	1	655.536
2	90.8102	2	232.7012	2	374.592	2	516.483	2	658.374
3	93.6480	3	235.5390	3	377.430	3	519.321	3	661.212
4	96.4858	4	238.3768	4	380.268	4	522.159	4	664.050
5	99.3237	5	241.2146	5	383.106	5	524.996	5	666.887
6	102.1615	6	244.0524	6	385.943	6	527.834	6	669.725
7	104.9993	7	246.8902	7	388.781	7	530.672	7	672.563
8	107.8371	8	249.7281	8	391.619	8	533.510	8	675.401
9	110.6749	9	252.5659	9	394.457	9	536.348	9	678.239
40	113.5128	90	255.4037	140	397.295	190	539.186	240	681.077
1	116.3506	1	258.2415	1	400.132	1	542.023	1	683.914
2	119.1884	2	261.0793	2	402.970	2	544.861	2	686.752
3	122.0262	3	263.9172	3	405.808	3	547.699	3	689.590
4	124.8640	4	266.7550	4	408.646	4	550.537	4	692.428
5	127.7019	5	269.5928	5	411.484	5	553.375	5	695.266
6	130.5397	6	272.4306	6	414.322	6	556.213	6	698.103
7	133.3775	7	275.2684	7	417.159	7	559.050	7	700.941
8	136.2153	8	278.1063	8	419.997	8	561.888	8	703.779
9	139.0531	9	280.9441	9	422.835	9	564.726	9	706.617

CAPACITY—HECTOLITRES TO BUSHELS

[Reduction factor: 1 hectolitre = 2.8378189 bushels]

Hecto-litres	Bushels	Hecto-litres	Bushels	Hecto-litres	Bushels	Hecto-litres	Bushels	Hecto-litres	Bushels
250	709.455	300	851.346	350	993.237	400	1,135.128	450	1,277.019
1	712.293	1	854.183	1	996.074	1	1,137.965	1	1,279.856
2	715.130	2	857.021	2	998.912	2	1,140.803	2	1,282.694
3	717.968	3	859.859	3	1,001.750	3	1,143.641	3	1,285.532
4	720.806	4	862.697	4	1,004.588	4	1,146.479	4	1,288.370
5	723.644	5	865.535	5	1,007.426	5	1,149.317	5	1,291.208
6	726.482	6	868.373	6	1,010.264	6	1,152.154	6	1,294.045
7	729.319	7	871.210	7	1,013.101	7	1,154.992	7	1,296.883
8	732.157	8	874.048	8	1,015.939	8	1,157.830	8	1,299.721
9	734.995	9	876.886	9	1,018.777	9	1,160.668	9	1,302.559
260	737.833	310	879.724	360	1,021.615	410	1,163.506	460	1,305.397
1	740.671	1	882.562	1	1,024.453	1	1,166.344	1	1,308.235
2	743.509	2	885.400	1	1,027.290	2	1,169.181	2	1,311.072
3	746.346	3	888.237	3	1,030.128	3	1,172.019	3	1,313.910
4	749.184	4	891.075	4	1,032.966	4	1,174.857	4	1,316.748
5	752.022	5	893.913	5	1,035.804	5	1,177.695	5	1,319.586
6	754.860	6	896.751	6	1,038.642	6	1,180.533	6	1,322.424
7	757.698	7	899.589	7	1,041.480	7	1,183.370	7	1,325.261
8	760.535	8	902.426	8	1,044.317	8	1,186.208	8	1,328.099
9	763.373	9	905.264	9	1,047.155	9	1,189.046	9	1,330.937
270	766.211	320	908.102	370	1,049.993	420	1,191.884	470	1,333.775
1	769.049	1	910.940	1	1,052.831	1	1,194.722	1	1,336.613
2	771.887	2	913.778	2	1,055.669	2	1,197.560	2	1,339.451
3	774.725	3	916.616	3	1,058.506	3	1,200.397	3	1,342.288
4	777.562	4	919.453	4	1,061.344	4	1,203.235	4	1,345.126
5	780.400	5	922.291	5	1,064.182	5	1,206.073	5	1,347.964
6	783.238	6	925.129	6	1,067.020	6	1,208.911	6	1,350.802
7	786.076	7	927.967	7	1,069.858	7	1,211.749	7	1,353.640
8	788.914	8	930.805	8	1,072.696	8	1,214.586	8	1,356.477
9	791.751	9	933.642	9	1,075.533	9	1,217.424	9	1,359.315
280	794.589	330	936.480	380	1,078.371	430	1,220.262	480	1,362.153
1	797.427	1	939.318	1	1,081.209	1	1,223.100	1	1,364.991
2	800.265	2	942.156	2	1,084.047	2	1,225.938	2	1,367.829
3	803.103	3	944.994	3	1,086.885	3	1,228.776	3	1,370.667
4	805.941	4	947.832	4	1,089.722	4	1,231.613	4	1,373.504
5	808.778	5	950.669	5	1,092.560	5	1,234.451	5	1,376.342
6	811.616	6	953.507	6	1,095.398	6	1,237.289	6	1,379.180
7	814.454	7	956.345	7	1,098.236	7	1,240.127	7	1,382.018
8	817.292	8	959.183	8	1,101.074	8	1,242.965	8	1,384.856
9	820.130	9	962.021	9	1,103.912	9	1,245.803	9	1,387.693
290	822.967	340	964.858	390	1,106.749	440	1,248.640	490	1,390.531
1	825.805	1	967.696	1	1,109.587	1	1,251.478	1	1,393.369
2	828.643	2	970.534	2	1,112.425	2	1,254.316	2	1,396.207
3	831.481	3	973.372	3	1,115.263	3	1,257.154	3	1,399.045
4	834.319	4	976.210	4	1,118.101	4	1,259.992	4	1,401.883
5	837.157	5	979.048	5	1,120.938	5	1,262.829	5	1,404.720
6	839.994	6	981.885	6	1,123.776	6	1,265.667	6	1,407.558
7	842.832	7	984.723	7	1,126.614	7	1,268.505	7	1,410.396
8	845.670	8	987.561	8	1,129.452	8	1,271.343	8	1,413.234
9	848.508	9	990.399	9	1,132.290	9	1,274.181	9	1,416.072

CAPACITY—HECTOLITRES TO BUSHELS

[Reduction factor: 1 hectolitre = 2.8378189 bushels]

Hecto-litres	Bushels	Hecto-litres	Bushels	Hecto-litres	Bushels	Hecto-litres	Bushels	Hecto-litres	Bushels
500	1,418.909	550	1,560.800	600	1,702.691	650	1,844.582	700	1,986.473
1	1,421.747	1	1,563.638	1	1,705.529	1	1,847.420	1	1,989.311
2	1,424.585	2	1,566.476	2	1,708.367	2	1,850.258	2	1,992.149
3	1,427.423	3	1,569.314	3	1,711.205	3	1,853.096	3	1,994.987
4	1,430.261	4	1,572.152	4	1,714.043	4	1,855.934	4	1,997.825
5	1,433.099	5	1,574.989	5	1,716.880	5	1,858.771	5	2,000.662
6	1,435.936	6	1,577.827	6	1,719.718	6	1,861.609	6	2,003.500
7	1,438.774	7	1,580.665	7	1,722.556	7	1,864.447	7	2,006.338
8	1,441.612	8	1,583.503	8	1,725.394	8	1,867.285	8	2,009.176
9	1,444.450	9	1,586.341	9	1,728.232	9	1,870.123	9	2,012.014
510	1,447.288	560	1,589.179	610	1,731.070	660	1,872.960	710	2,014.851
1	1,450.125	1	1,592.016	1	1,733.907	1	1,875.798	1	2,017.689
2	1,452.963	2	1,594.854	2	1,736.745	2	1,878.636	2	2,020.527
3	1,455.801	3	1,597.692	3	1,739.583	3	1,881.474	3	2,023.365
4	1,458.639	4	1,600.530	4	1,742.421	4	1,884.312	4	2,026.203
5	1,461.477	5	1,603.368	5	1,745.259	5	1,887.150	5	2,029.041
6	1,464.315	6	1,606.206	6	1,748.096	6	1,889.987	6	2,031.878
7	1,467.152	7	1,609.043	7	1,750.934	7	1,892.825	7	2,034.716
8	1,469.990	8	1,611.881	8	1,753.772	8	1,895.663	8	2,037.554
9	1,472.828	9	1,614.719	9	1,756.610	9	1,898.501	9	2,040.392
520	1,475.666	570	1,617.557	620	1,759.448	670	1,901.339	720	2,043.230
1	1,478.504	1	1,620.395	1	1,762.286	1	1,904.176	1	2,046.067
2	1,481.341	2	1,623.232	2	1,765.123	2	1,907.014	2	2,C48.905
3	1,484.179	3	1,626.070	3	1,767.961	3	1,909.852	3	2,051.743
4	1,487.017	4	1,628.908	4	1,770.799	4	1,912.690	4	2,054.581
5	1,489.855	5	1,631.746	5	1,773.637	5	1,915.528	5	2,057.419
6	1,492.693	6	1,634.584	6	1,776.475	6	1,918.366	6	2,060.257
7	1,495.531	7	1,637.422	7	1,779.312	7	1,921.203	7	2,063.094
8	1,498.368	8	1,640.259	8	1,782.150	8	1,924.041	8	2,065.932
9	1,501.206	9	1,643.097	9	1,784.988	9	1,926.879	9	2,068.770
530	1,504.044	580	1,645.935	630	1,787.826	680	1,929.717	730	2,071.608
1	1,506.882	1	1,648.773	1	1,790.664	1	1,932.555	1	2,074.446
2	1,509.720	2	1,651.611	2	1,793.502	2	1,935.393	2	2,077.283
3	1,512.557	3	1,654.448	3	1,796.339	3	1,938.230	3	2,080.121
4	1,515.395	4	1,657.286	4	1,799.177	4	1,941.068	4	2,082.959
5	1,518.233	5	1,660.124	5	1,802.015	5	1,943.906	5	2,085.797
6	1,521.071	6	1,662.962	6	1,804.853	6	1,946.744	6	2,088.635
7	1,523.909	7	1,665.800	7	1,807.691	7	1,949.582	7	3,091.473
8	1,526.747	8	1,668.638	8	1,810.528	8	1,952.419	8	2,094.310
9	1,529.584	9	1,671.475	9	1,813.366	9	1,955.257	9	2,097.148
540	1,532.422	590	1,674.313	640	1,816.204	690	1,958.095	740	2,099.986
1	1,535.260	1	1,677.151	1	1,819.042	1	1,960.933	1	2,102.824
2	1,538.098	2	1,679.989	2	1,821.880	2	1,963.771	2	2,105.662
3	1,540.936	3	1,682.827	3	1,824.718	3	1,966.609	3	2,108.499
4	1,543.773	4	1,685.664	4	1,827.555	4	1,969.446	4	2,111.337
5	1,546.611	5	1,688.502	5	1,830.393	5	1,972.284	5	2,114.175
6	1,549.449	6	1,691.340	6	1,833.231	6	1,975.122	6	2,117.013
7	1,552.287	7	1,694.178	7	1,836.069	7	1,977.960	7	2,119.851
8	1,555.125	8	1,697.016	8	1,838.907	8	1,980.798	8	2,122.689
9	1,557.963	9	1,699.854	9	1,841.744	9	1,983.635	9	2,125.526

CAPACITY—HECTOLITRES TO BUSHELS

[Reduction factor: 1 hectolitre = 2.8378189 bushels]

Hecto-litres	Bush-els	Hecto-litres	Bush-els	Hecto-litres	Bush-els	Hecto-litres	Bush-els	Hecto-litres	Bush-els
750	2,128. 364	800	2,270. 255	850	2,412. 146	900	2,554. 037	950	2,695. 928
1	2,131. 202	1	2,273. 093	1	2,414. 984	1	2,556. 875	1	2,698. 766
2	2,134. 040	2	2,275. 931	2	2,417. 822	2	2,559. 713	2	2,701. 604
3	2,136. 878	3	2,278. 769	3	2,420. 660	3	2,562. 550	3	2,704. 441
4	2,139. 715	4	2,281. 606	4	2,423. 497	4	2,565. 388	4	2,707. 279
5	2,142. 553	5	2,284. 444	5	2,426. 335	5	2,568. 226	5	2,710. 117
6	2,145. 391	6	2,287. 282	6	2,429. 173	6	2,571. 064	6	2,712. 955
7	2,148. 229	7	2,290. 120	7	2,432. 011	7	2,573. 902	7	2,715. 793
8	2,151. 067	8	2,292. 958	8	2,434. 849	8	2,576. 740	8	2,718. 631
9	2,153. 905	9	2,295. 796	9	2,437. 686	9	2,579. 577	9	2,721. 468
760	2,156. 742	810	2,298. 633	860	2,440. 524	910	2,582. 415	960	2,724. 306
1	2,159. 580	1	2,301. 471	1	2,443. 362	1	2,585. 253	1	2,727. 144
2	2,162. 418	2	2,304. 309	2	2,446. 200	2	2,588. 091	2	2,729. 982
3	2,165. 256	3	2,307. 147	3	2,449. 038	3	2,590. 929	3	2,732. 820
4	2,168. 094	4	2,309. 985	4	2,451. 876	4	2,593. 766	4	2,735. 657
5	2,170. 931	5	2,312. 822	5	2,454. 713	5	2,596. 604	5	2,738. 495
6	2,173. 769	6	2,315. 660	6	2,457. 551	6	2,599. 442	6	2,741. 333
7	2,176. 607	7	2,318. 498	7	2,460. 389	7	2,602. 280	7	2,744. 171
8	2,179. 445	8	2,321. 336	8	2,463. 227	8	2,605. 118	8	2,747. 009
9	2,182. 283	9	2,324. 174	9	2,466. 065	9	2,607. 956	9	2,749. 847
770	2,185. 121	820	2,327. 012	870	2,468. 902	920	2,610. 793	970	2,752. 684
1	2,187. 958	1	2,329. 849	1	2,471. 740	1	2,613. 631	1	2,755. 522
2	2,190. 796	2	2,332. 687	2	2,474. 578	2	2,616. 469	2	2,758. 360
3	2,193. 634	3	2,335. 525	3	2,477. 416	3	2,619. 307	3	2,761. 198
4	2,196. 472	4	2,338. 363	4	2,480. 254	4	2,622. 145	4	2,764. 036
5	2,199. 310	5	2,341. 201	5	2,483. 092	5	2,624. 982	5	2,766. 873
6	2,202. 147	6	2,344. 038	6	2,485. 929	6	2,627. 820	6	2,769. 711
7	2,204. 985	7	2,346. 876	7	2,488. 767	7	2,630. 658	7	2,772. 549
8	2,207. 823	8	2,349. 714	8	2,491. 605	8	2,633. 496	8	2,775. 387
9	2,210. 661	9	2,352. 552	9	2,494. 443	9	2,636. 334	9	2,778. 225
780	2,213. 499	830	2,355. 390	880	2,497. 281	930	2,639. 172	980	2,781. 063
1	2,216. 337	1	2,358. 228	1	2,500. 118	1	2,642. 009	1	2,783. 900
2	2,219. 174	2	2,361. 065	2	2,502. 956	2	2,644. 847	2	2,786. 738
3	2,222. 012	3	2,363. 903	3	2,505. 794	3	2,647. 685	3	2,789. 576
4	2,224. 850	4	2,366. 741	4	2,508. 632	4	2,650. 523	4	2,792. 414
5	2,227. 688	5	2,369. 579	5	2,511. 470	5	2,653. 361	5	2,795. 252
6	2,230. 526	6	2,372. 417	6	2,514. 308	6	2,656. 199	6	2,798. 089
7	2,233. 363	7	2,375. 254	7	2,517. 145	7	2,659. 036	7	2,800. 927
8	2,236. 201	8	2,378. 092	8	2,519. 983	8	2,661. 874	8	2,803. 765
9	2,239. 039	9	2,380. 930	9	2,522. 821	9	2,664. 712	9	2,806. 603
790	2,241. 877	840	2,383. 768	890	2,525. 659	940	2,667. 550	990	2,809. 441
1	2,244. 715	1	2,386. 606	1	2,528. 497	1	2,670. 388	1	2,812. 279
2	2,247. 553	2	2,389. 444	2	2,531. 334	2	2,673. 225	2	2,815. 116
3	2,250. 390	3	2,392. 281	3	2,534. 172	3	2,676. 063	3	2,817. 954
4	2,253. 228	4	2,395. 119	4	2,537. 010	4	2,678. 901	4	2,820. 792
5	2,256. 066	5	2,397. 957	5	2,539. 848	5	2,681. 739	5	2,823. 630
6	2,258. 904	6	2,400. 795	6	2,542. 686	6	2,684. 577	6	2,826. 468
7	2,261. 742	7	2,403. 633	7	2,545. 524	7	2,687. 415	7	2,829. 305
8	2,264. 579	8	2,406. 470	8	2,548. 361	8	2,690. 252	8	2,832. 143
9	2,267. 417	9	2,409. 308	9	2,551. 199	9	2,693. 090	9	2,834. 981

CAPACITY—BUSHELS TO HECTOLITRES

[Reduction factor: 1 bushel = 0.35238330 hectolitre]

Bush-els	Hecto-litres	Bush-els	Hecto-litres	Bush-els	Hecto-litres	Bush-els	Hecto-litres	Bush-els	Hecto-litres
0		50	17.61917	100	35.2383	150	52.8575	200	70.4767
1	0.35238	1	17.97155	1	35.5907	1	53.2099	1	70.8290
2	0.70477	2	18.32393	2	35.9431	2	53.5623	2	71.1814
3	1.05715	3	18.67631	3	36.2955	3	53.9146	3	71.5338
4	1.40953	4	19.02870	4	36.6479	4	54.2670	4	71.8862
5	1.76192	5	19.38108	5	37.0002	5	54.6194	5	72.2386
6	2.11430	6	19.73346	6	37.3526	6	54.9718	6	72.5910
7	2.46668	7	20.08585	7	37.7050	7	55.3242	7	72.9433
8	2.81907	8	20.43823	8	38.0574	8	55.6766	8	73.2957
9	3.17145	9	20.79061	9	38.4098	9	56.0289	9	73.6481
10	3.52383	60	21.14300	110	38.7622	160	56.3813	210	74.0005
1	3.87622	1	21.49538	1	39.1145	1	56.7337	1	74.3529
2	4.22860	2	21.84776	2	39.4669	2	57.0861	2	74.7053
3	4.58098	3	22.20015	3	39.8193	3	57.4385	3	75.0576
4	4.93337	4	22.55253	4	40.1717	4	57.7909	4	75.4100
5	5.28575	5	22.90491	5	40.5241	5	58.1432	5	75.7624
6	5.63813	6	23.25730	6	40.8765	6	58.4956	6	76.1148
7	5.99052	7	23.60968	7	41.2288	7	58.8480	7	76.4672
8	6.34290	8	23.96206	8	41.5812	8	59.2004	8	76.8196
9	6.69528	9	24.31445	9	41.9336	9	59.5528	9	77.1719
20	7.04767	70	24.66683	120	42.2860	170	59.9052	220	77.5243
1	7.40005	1	25.01921	1	42.6384	1	60.2575	1	77.8767
2	7.75243	2	25.37160	2	42.9908	2	60.6099	2	78.2291
3	8.10482	3	25.72398	3	43.3431	3	60.9623	3	78.5815
4	8.45720	4	26.07636	4	43.6955	4	61.3147	4	78.9339
5	8.80958	5	26.42875	5	44.0479	5	61.6671	5	79.2862
6	9.16197	6	26.78113	6	44.4003	6	62.0195	6	79.6386
7	9.51435	7	27.13351	7	44.7527	7	62.3718	7	79.9910
8	9.86673	8	27.48590	8	45.1051	8	62.7242	8	80.3434
9	10.21912	9	27.83828	9	45.4574	9	63.0766	9	80.6958
30	10.57150	80	28.19066	130	45.8098	180	63.4290	230	81.0482
1	10.92388	1	28.54305	1	46.1622	1	63.7814	1	81.4005
2	11.27627	2	28.89543	2	46.5146	2	64.1338	2	81.7529
3	11.62865	3	29.24781	3	46.8670	3	64.4861	3	82.1053
4	11.98103	4	29.60020	4	47.2194	4	64.8385	4	82.4577
5	12.33342	5	29.95258	5	47.5717	5	65.1909	5	82.8101
6	12.68580	6	30.30496	6	47.9241	6	65.5433	6	83.1625
7	13.03818	7	30.65735	7	48.2765	7	65.8957	7	83.5148
8	13.39057	8	31.00973	8	48.6289	8	66.2481	8	83.8672
9	13.74295	9	31.36211	9	48.9813	9	66.6004	9	84.2196
40	14.09533	90	31.71450	140	49.3337	190	66.9528	240	84.5720
1	14.44772	1	32.06688	1	49.6860	1	67.3052	1	84.9244
2	14.80010	2	32.41926	2	50.0384	2	67.6576	2	85.2768
3	15.15248	3	32.77165	3	50.3908	3	68.0100	3	85.6291
4	15.50487	4	33.12403	4	50.7432	4	68.3624	4	85.9815
5	15.85725	5	33.47641	5	51.0956	5	68.7147	5	86.3339
6	16.20963	6	33.82880	6	51.4480	6	69.0671	6	86.6863
7	16.56202	7	34.18118	7	51.8003	7	69.4195	7	87.0387
8	16.91440	8	34.53356	8	52.1527	8	69.7719	8	87.3911
9	17.26678	9	34.88595	9	52.5051	9	70.1243	9	87.7434

CAPACITY—BUSHELS TO HECTOLITRES

[Reduction factor: 1 bushel = 0.35238330 hectolitre]

Bush-els	Hecto-litres	Bush-els	Hecto-litres	Bush-els	Hecto-litres	Bush-els	Hecto-litres	Bush-els	Hecto-litres
250	88.0958	300	105.7150	350	123.3342	400	140.9533	450	158.5725
1	88.4482	1	106.0674	1	123.6865	1	141.3057	1	158.9249
2	88.8006	2	106.4198	2	124.0389	2	141.6581	2	159.2773
3	89.1530	3	106.7721	3	124.3913	3	142.0105	3	159.6296
4	89.5054	4	107.1245	4	124.7437	4	142.3629	4	159.9820
5	89.8577	5	107.4769	5	125.0961	5	142.7152	5	160.3344
6	90.2101	6	107.8293	6	125.4485	6	143.0676	6	160.6868
7	90.5625	7	108.1817	7	125.8008	7	143.4200	7	161.0392
8	90.9149	8	108.5341	8	126.1532	8	143.7724	8	161.3916
9	91.2673	9	108.8864	9	126.5056	9	144.1248	9	161.7439
260	91.6197	310	109.2388	360	126.8580	410	144.4772	460	162.0963
1	91.9720	1	109.5912	1	127.2104	1	144.8295	1	162.4487
2	92.3244	2	109.9436	2	127.5628	2	145.1819	2	162.8011
3	92.6768	3	110.2960	3	127.9151	3	145.5343	3	163.1535
4	93.0292	4	110.6484	4	128.2675	4	145.8867	4	163.5059
5	93.3816	5	111.0007	5	128.6199	5	146.2391	5	163.8582
6	93.7340	6	111.3531	6	128.9723	6	146.5915	6	164.2106
7	94.0863	7	111.7055	7	129.3247	7	146.9438	7	164.5630
8	94.4387	8	112.0579	8	129.6771	8	147.2962	8	164.9154
9	94.7911	9	112.4103	9	130.0294	9	147.6486	9	165.2678
270	95.1435	320	112.7627	370	130.3818	420	148.0010	470	165.6202
1	95.4959	1	113.1150	1	130.7342	1	148.3534	1	165.9725
2	95.8483	2	113.4674	2	131.0866	2	148.7058	2	166.3249
3	96.2006	3	113.8198	3	131.4390	3	149.0581	3	166.6773
4	96.5530	4	114.1722	4	131.7914	4	149.4105	4	167.0297
5	96.9054	5	114.5746	5	132.1437	5	149.7629	5	167.3821
6	97.2578	6	114.8770	6	132.4961	6	150.1153	6	167.7345
7	97.6102	7	115.2293	7	132.8485	7	150.4677	7	168.0868
8	97.9626	8	115.5817	8	133.2009	8	150.8201	8	168.4392
9	98.3149	9	115.9341	9	133.5533	9	151.1724	9	168.7916
280	98.6673	330	116.2865	380	133.9057	430	151.5248	480	169.1440
1	99.0197	1	116.6389	1	134.2580	1	151.8772	1	169.4964
2	99.3721	2	116.9913	2	134.6104	2	152.2296	2	169.8488
3	99.7245	3	117.3436	3	134.9628	3	152.5820	3	170.2011
4	100.0769	4	117.6960	4	135.3152	4	152.9344	4	170.5535
5	100.4292	5	118.0484	5	135.6676	5	153.2867	5	170.9059
6	100.7816	6	118.4008	6	136.0200	6	153.6391	6	171.2583
7	101.1340	7	118.7532	7	136.3723	7	153.9915	7	171.6107
8	101.4864	8	119.1056	8	136.7247	8	154.3439	8	171.9631
9	101.8388	9	119.4579	9	137.0771	9	154.6963	9	172.3154
290	102.1912	340	119.8103	390	137.4295	440	155.0487	490	172.6678
1	102.5435	1	120.1627	1	137.7819	1	155.4010	1	173.0202
2	102.8959	2	120.5151	2	138.1343	2	155.7534	2	173.3726
3	103.2483	3	120.8675	3	138.4866	3	156.1058	3	173.7250
4	103.6007	4	121.2199	4	138.8390	4	156.4582	4	174.0774
5	103.9531	5	121.5722	5	139.1914	5	156.8106	5	174.4297
6	104.3055	6	121.9246	6	139.5438	6	157.1630	6	174.7821
7	104.6578	7	122.2770	7	139.8962	7	157.5153	7	175.1345
8	105.0102	8	122.6294	8	140.2486	8	157.8677	8	175.4869
9	105.3626	9	122.9818	9	140.6009	9	158.2201	9	175.8393

CAPACITY—BUSHELS TO HECTOLITRES

[Reduction factor: 1 bushel = 0.35238330 hectolitre]

Bush-els	Hecto-litres	Bush-els	Hecto-litres	Bush-els	Hecto-litres	Bush-els	Hecto-litres	Bush-els	Hecto-litres
500	176.1917	550	193.8108	600	211.4300	650	229.0491	700	246.6683
1	176.5440	1	194.1632	1	211.7824	1	229.4015	1	247.0207
2	176.8964	2	194.5156	2	212.1347	2	229.7539	2	247.3731
3	177.2488	3	194.8680	3	212.4871	3	230.1063	3	247.7255
4	177.6012	4	195.2203	4	212.8395	4	230.4587	4	248.0778
5	177.9536	5	195.5727	5	213.1919	5	230.8111	5	248.4302
6	178.3060	6	195.9251	6	213.5443	6	231.1634	6	248.7826
7	178.6583	7	196.2775	7	213.8967	7	231.5158	7	249.1350
8	179.0107	8	196.6299	8	214.2490	8	231.8682	8	249.4874
9	179.3631	9	196.9823	9	214.6014	9	232.2206	9	249.8398
510	179.7155	560	197.3346	610	214.9538	660	232.5730	710	250.1921
1	180.0679	1	197.6870	1	215.3062	1	232.9254	1	250.5445
2	180.4203	2	198.0394	2	215.6586	2	233.2777	2	250.8969
3	180.7726	3	198.3918	3	216.0110	3	233.6301	3	251.2493
4	181.1250	4	198.7442	4	216.3633	4	233.9825	4	251.6017
5	181.4774	5	199.0966	5	216.7157	5	234.3349	5	251.9541
6	181.8298	6	199.4489	6	217.0681	6	234.6873	6	252.3064
7	182.1822	7	199.8013	7	217.4205	7	235.0397	7	252.6588
8	182.5346	8	200.1537	8	217.7729	8	235.3920	8	253.0112
9	182.8869	9	200.5061	9	218.1253	9	235.7444	9	253.3636
520	183.2393	570	200.8585	620	218.4776	670	236.0968	720	253.7160
1	183.5917	1	201.2109	1	218.8300	1	236.4492	1	254.0684
2	183.9441	2	201.5632	2	219.1824	2	236.8016	2	254.4207
3	184.2965	3	201.9156	3	219.5348	3	237.1540	3	254.7731
4	184.6489	4	202.2680	4	219.8872	4	237.5063	4	255.1255
5	185.0012	5	202.6204	5	220.2396	5	237.8587	5	255.4779
6	185.3536	6	202.9728	6	220.5919	6	238.2111	6	255.8303
7	185.7060	7	203.3252	7	220.9443	7	238.5635	7	256.1827
8	186.0584	8	203.6775	8	221.2967	8	238.9159	8	256.5350
9	186.4108	9	204.0299	9	221.6491	9	239.2683	9	256.8874
530	186.7631	580	204.3823	630	222.0015	680	239.6206	730	257.2398
1	187.1155	1	204.7347	1	222.3539	1	239.9730	1	257.5922
2	187.4679	2	205.0871	2	222.7062	2	240.3254	2	257.9446
3	187.8203	3	205.4395	3	223.0586	3	240.6778	3	258.2970
4	188.1727	4	205.7918	4	223.4110	4	241.0302	4	258.6493
5	188.5251	5	206.1442	5	223.7634	5	241.3826	5	259.0017
6	188.8774	6	206.4966	6	224.1158	6	241.7349	6	259.3541
7	189.2298	7	206.8490	7	224.4682	7	242.0873	7	259.7065
8	189.5822	8	207.2014	8	224.8205	8	242.4397	8	260.0589
9	189.9346	9	207.5538	9	225.1729	9	242.7921	9	260.4113
540	190.2870	590	207.9061	640	225.5253	690	243.1445	740	260.7636
1	190.6394	1	208.2585	1	225.8777	1	243.4969	1	261.1160
2	190.9917	2	208.6109	2	226.2301	2	243.8492	2	261.4684
3	191.3441	3	208.9633	3	226.5825	3	244.2016	3	261.8208
4	191.6965	4	209.3157	4	226.9348	4	244.5540	4	262.1732
5	192.0489	5	209.6681	5	227.2872	5	244.9064	5	262.5256
6	192.4013	6	210.0204	6	227.6396	6	245.2588	6	262.8779
7	192.7537	7	210.3728	7	227.9920	7	245.6112	7	263.2303
8	193.1060	8	210.7252	8	228.3444	8	245.9635	8	263.5827
9	193.4584	9	211.0776	9	228.6968	9	246.3159	9	263.9351

CAPACITY—BUSHELS TO HECTOLITRES

[Reduction factor: 1 bushel = 0.35238330 hectolitre]

Bushels	Hectolitres	Bushels	Hectolitres	Bushels	Hectolitres	Bushels	Hectolitres	Bushels	Hectolitres
750	264.2875	800	281.9066	850	299.5258	900	317.1450	950	334.7641
1	264.6399	1	282.2590	1	299.8782	1	317.4974	1	335.1165
2	264.9922	2	282.6114	2	300.2306	2	317.8497	2	335.4689
3	265.3446	3	282.9638	3	300.5830	3	318.2021	3	335.8213
4	265.6970	4	283.3162	4	300.9353	4	318.5545	4	336.1737
5	266.0494	5	283.6686	5	301.2877	5	318.9069	5	336.5261
6	266.4018	6	284.0209	6	301.6401	6	319.2593	6	336.8784
7	266.7542	7	284.3733	7	301.9925	7	319.6117	7	337.2308
8	267.1065	8	284.7257	8	302.3449	8	319.9640	8	337.5832
9	267.4589	9	285.0781	9	302.6973	9	320.3164	9	337.9356
760	267.8113	810	285.4305	860	303.0496	910	320.6688	960	338.2880
1	268.1637	1	285.7829	1	303.4020	1	321.0212	1	338.6404
2	268.5161	2	286.1352	2	303.7544	2	321.3736	2	338.9927
3	268.8685	3	286.4876	3	304.1068	3	231.7260	3	339.3451
4	269.2208	4	286.8400	4	304.4592	4	322.0783	4	339.6975
5	269.5732	5	287.1924	5	304.8116	5	322.4307	5	340.0499
6	269.9256	6	287.5448	6	305.1639	6	322.7831	6	340.4023
7	270.2780	7	287.8972	7	305.5163	7	323.1355	7	340.7547
8	270.6304	8	288.2495	8	305.8687	8	323.4879	8	341.1070
9	270.9828	9	288.6019	9	306.2211	9	323.8403	9	341.4594
770	271.3351	820	288.9543	870	306.5735	920	324.1926	970	341.8118
1	271.6875	1	289.3067	1	306.9259	1	324.5450	1	342.1642
2	272.0399	2	289.6591	2	307.2782	2	324.8974	2	342.5166
3	272.3923	3	290.0115	3	307.6306	3	325.2498	3	342.8690
4	272.7447	4	290.3638	4	307.9830	4	325.6022	4	343.2213
5	273.0971	5	290.7162	5	308.3354	5	325.9546	5	343.5737
6	273.4494	6	291.0686	6	308.6878	6	326.3069	6	343.9261
7	273.8018	7	291.4210	7	309.0402	7	326.6593	7	344.2785
8	274.1542	8	291.7734	8	309.3925	8	327.0117	8	344.6309
9	274.5066	9	292.1258	9	309.7449	9	327.3641	9	344.9833
780	274.8590	830	292.4781	880	310.0973	930	327.7165	980	345.3356
1	275.2114	1	292.8305	1	310.4497	1	328.0689	1	345.6880
2	275.5637	2	293.1829	2	310.8021	2	328.4212	2	346.0404
3	275.9161	3	293.5353	3	311.1545	3	328.7736	3	346.3928
4	276.2685	4	293.8877	4	311.5068	4	329.1260	4	346.7452
5	276.6209	5	294.2401	5	311.8592	5	329.4784	5	347.0976
6	276.9733	6	294.5924	6	312.2116	6	329.8308	6	347.4499
7	277.3257	7	294.9448	7	312.5640	7	330.1832	7	347.8023
8	277.6780	8	295.2972	8	312.9164	8	330.5355	8	348.1547
9	278.0304	9	295.6496	9	313.2688	9	330.8879	9	348.5071
790	278.3828	840	296.0020	890	313.6211	940	331.2403	990	348.8595
1	278.7352	1	296.3544	1	313.9735	1	331.5927	1	349.2119
2	279.0876	2	296.7067	2	314.3259	2	331.9451	2	349.5642
3	279.4400	3	297.0591	3	314.6783	3	332.2975	3	349.9166
4	279.7923	4	297.4115	4	315.0307	4	332.6498	4	350.2690
5	280.1447	5	297.7639	5	315.3831	5	333.0022	5	350.6214
6	280.4971	6	298.1163	6	315.7354	6	333.3546	6	350.9738
7	280.8495	7	298.4687	7	316.0878	7	333.7070	7	351.3262
8	281.2019	8	298.8210	8	316.4402	8	334.0594	8	351.6785
9	281.5543	9	299.1734	9	316.7926	9	334.4118	9	352.0309

MASS—KILOGRAMS TO AVOIRDUPOIS POUNDS

[Reduction factor: 1 kilogram = 2.204622341 avoirdupois pounds]

Kilos	Pounds	Kilos	Pounds	Kilos	Pounds	Kilos	Pounds	Kilos	Pounds
0		50	110.2311	100	220.4622	150	330.6934	200	440.9245
1	2.2046	1	112.4357	1	222.6669	1	332.8980	1	443.1291
2	4.4092	2	114.6404	2	224.8715	2	335.1026	2	445.3337
3	6.6139	3	116.8450	3	227.0761	3	337.3072	3	447.5383
4	8.8185	4	119.0496	4	229.2807	4	339.5118	4	449.7430
5	11.0231	5	121.2542	5	231.4853	5	341.7165	5	451.9476
6	13.2277	6	123.4589	6	233.6900	6	343.9211	6	454.1522
7	15.4324	7	125.6635	7	235.8946	7	346.1257	7	456.3568
8	17.6370	8	127.8681	8	238.0992	8	348.3303	8	458.5614
9	19.8416	9	130.0727	9	240.3038	9	350.5350	9	460.7661
10	22.0462	60	132.2773	110	242.5085	160	352.7396	210	462.9707
1	24.2508	1	134.4820	1	244.7131	1	354.9442	1	465.1753
2	26.4555	2	136.6866	2	246.9177	2	357.1488	2	467.3799
3	28.6601	3	138.8912	3	249.1223	3	359.3534	3	469.5846
4	30.8647	4	141.0958	4	251.3269	4	361.5581	4	471.7892
5	33.0693	5	143.3005	5	253.5316	5	363.7627	5	473.9938
6	35.2740	6	145.5051	6	255.7362	6	365.9673	6	476.1984
7	37.4786	7	147.7097	7	257.9408	7	368.1719	7	478.4030
8	39.6832	8	149.9143	8	260.1454	8	370.3766	8	480.6077
9	41.8878	9	.152.1189	9	262.3501	9	372.5812	9	482.8123
20	44.0924	70	154.3236	120	264.5547	170	374.7858	220	485.0169
1	46.2971	1	156.5282	1	266.7593	1	376.9904	1	487.2215
2	48.5017	2	158.7328	2	268.9639	2	379.1950	2	489.4262
3	50.7063	3	160.9374	3	271.1685	3	381.3997	3	491.6308
4	52.9109	4	163.1421	4	273.3732	4	383.6043	4	493.8354
5	55.1156	5	165.3467	5	275.5778	5	385.8089	5	496.0400
6	57.3202	6	167.5513	6	277.7824	6	388.0135	6	498.2446
7	59.5248	7	169.7559	7	279.9870	7	390.2182	7	500.4493
8	61.7294	8	171.9605	8	282.1917	8	392.4228	8	502.6539
9	63.9340	9	174.1652	9	284.3963	9	394.6274	9	504.8585
30	66.1387	80	176.3698	130	286.6009	180	396.8320	230	507.0631
1	68.3433	1	178.5744	1	288.8055	1	399.0366	1	509.2678
2	70.5479	2	180.7790	2	291.0101	2	401.2413	2	511.4724
3	72.7525	3	182.9837	3	293.2148	3	403.4459	3	513.6770
4	74.9572	4	185.1883	4	295.4194	4	405.6505	4	515.8816
5	77.1618	5	187.3929	5	297.6240	5	407.8551	5	518.0863
6	79.3664	6	189.5975	6	299.8286	6	410.0598	6	520.2909
7	81.5710	7	191.8021	7	302.0333	7	412.2644	7	522.4955
8	83.7756	8	194.0068	8	304.2379	8	414.4690	8	524.7001
9	85.9803	9	196.2114	9	306.4425	9	416.6736	9	526.9047
40	88.1849	90	198.4160	140	308.6471	190	418.8782	240	529.1094
1	90.3895	1	200.6206	1	310.8518	1	421.0829	1	531.3140
2	92.5941	2	202.8253	2	313.0564	2	423.2875	2	533.5186
3	94.7988	3	205.0299	3	315.2610	3	425.4921	3	535.7232
4	97.0034	4	207.2345	4	317.4656	4	427.6967	4	537.9279
5	99.2080	5	209.4391	5	319.6702	5	429.9014	5	540.1325
6	101.4126	6	211.6437	6	321.8749	6	432.1060	6	542.3371
7	103.6173	7	213.8484	7	324.0795	7	434.3106	7	544.5417
8	105.8219	8	216.0530	8	326.2841	8	436.5152	8	546.7463
9	108.0265	9	218.2576	9	328.4887	9	438.7198	9	548.9510

MASS—KILOGRAMS TO AVOIRDUPOIS POUNDS

[Reduction factor: 1 kilogram = 2.204622341 avoirdupois pounds]

Kilos	Pounds	Kilos	Pounds	Kilos	Pounds	Kilos	Pounds	Kilos	Pounds
250	551. 1556	300	661. 3867	350	771. 6178	400	881. 8489	450	992. 0801
1	553. 3602	1	663. 5913	1	773. 8224	1	884. 0536	1	994. 2847
2	555. 5648	2	665. 7959	2	776. 0271	2	886. 2582	2	996. 4893
3	557. 7695	3	668. 0006	3	778. 2317	3	888. 4628	3	998. 6939
4	559. 9741	4	670. 2052	4	780. 4363	4	890. 6674	4	1,000. 8985
5	562. 1787	5	672. 4098	5	782. 6409	5	892. 8720	5	1,003. 1032
6	564. 3833	6	674. 6144	6	784. 8456	6	895. 0767	6	1,005. 3078
7	566. 5879	7	676. 8191	7	787. 0502	7	897. 2813	7	1,007. 5124
8	568. 7926	8	679. 0237	8	789. 2548	8	899. 4859	8	1,009. 7170
9	570. 9972	9	681. 2283	9	791. 4594	9	901. 6905	9	1,011. 9217
260	573. 2018	310	683. 4329	360	793. 6640	410	903. 8952	460	1,014. 1263
1	575. 4064	1	685. 6375	1	795. 8687	1	906. 0998	1	1,016. 3309
2	577. 6111	2	687. 8422	2	798. 0733	2	908. 3044	2	1,018. 5355
3	579. 8157	3	690. 0468	3	800. 2779	3	910. 5090	3	1,020. 7401
4	582. 0203	4	692. 2514	4	802. 4825	4	912. 7136	4	1,022. 9448
5	584. 2249	5	694. 4560	5	804. 6872	5	914. 9183	5	1,025. 1494
6	586. 4295	6	696. 6607	6	806. 8918	6	917. 1229	6	1,027. 3540
7	588. 6342	7	698. 8653	7	809. 0964	7	919. 3275	7	1,029. 5586
8	590. 8388	8	701. 0699	8	811. 3010	8	921. 5321	8	1,031. 7633
9	593. 0434	9	703. 2745	9	813. 5056	9	923. 7368	9	1,033. 9679
270	595. 2480	320	705. 4791	370	815. 7103	420	925. 9414	470	1,036. 1725
1	597. 4527	1	707. 6838	1	817. 9149	1	928. 1460	1	1,038. 3771
2	599. 6573	2	709. 8884	2	820. 1195	2	930. 3506	2	1,040. 5817
3	601. 8619	3	712. 0930	3	822. 3241	3	932. 5553	3	1,042. 7864
4	604. 0665	4	714. 2976	4	824. 5288	4	934. 7599	4	1,044. 9910
5	606. 2711	5	716. 5023	5	826. 7334	5	936. 9645	5	1,047. 1956
6	608. 4758	6	718. 7069	6	828. 9380	6	939. 1691	6	1,049. 4002
7	610. 6804	7	720. 9115	7	831. 1426	7	941. 3737	7	1,051. 6049
8	612. 8850	8	723. 1161	8	833. 3472	8	943. 5784	8	1,053. 8095
9	615. 0896	9	725. 3208	9	835. 5519	9	945. 7830	9	1,056. 0141
280	617. 2943	330	727. 5254	380	837. 7565	430	947. 9876	480	1,058. 2187
1	619. 4989	1	729. 7300	1	839. 9611	1	950. 1922	1	1,060. 4233
2	621. 7035	2	731. 9346	2	842. 1657	2	952. 3969	2	1,062. 6280
3	623. 9081	3	734. 1392	3	844. 3704	3	954. 6015	3	1,064. 8326
4	626. 1127	4	736. 3439	4	846. 5750	4	956. 8061	4	1,067. 0372
5	628. 3174	5	738. 5485	5	848. 7796	5	959. 0107	5	1,069. 2418
6	630. 5220	6	740. 7531	6	850. 9842	6	961. 2153	6	1,071. 4465
7	632. 7266	7	742. 9577	7	853. 1888	7	963. 4200	7	1,073. 6511
8	634. 9312	8	745. 1624	8	855. 3935	8	965. 6246	8	1,075. 8557
9	637. 1359	9	747. 3670	9	857. 5981	9	967. 8292	9	1,078. 0603
290	639. 3405	340	749. 5716	390	859. 8027	440	970. 0338	490	1,080. 2649
1	641. 5451	1	751. 7762	1	862. 0073	1	972. 2385	1	1,082. 4696
2	643. 7497	2	753. 9808	2	864. 2120	2	974. 4431	2	1,084. 6742
3	645. 9543	3	756. 1855	3	866. 4166	3	976. 6477	3	1,086. 8788
4	648. 1590	4	758. 3901	4	868. 6212	4	978. 8523	4	1,089. 0834
5	650. 3636	5	760. 5947	5	870. 8258	5	981. 0569	5	1,091. 2881
6	652. 5682	6	762. 7993	6	873. 0304	6	983. 2616	6	1,093. 4927
7	654. 7728	7	765. 0040	7	875. 2351	7	985. 4662	7	1,095. 6973
8	656. 9775	8	767. 2086	8	877. 4397	8	987. 6708	8	1,097. 9019
9	659. 1821	9	769. 4132	9	879. 6443	9	989. 8754	9	1,100. 1065

MASS—KILOGRAMS TO AVOIRDUPOIS POUNDS

[Reduction factor: 1 kilogram = 2.204622341 avoirdupois pounds]

Kilos	Pounds	Kilos	Pounds	Kilos	Pounds	Kilos	Pounds	Kilos	Pounds
500	1,102. 3112	550	1,212. 5423	600	1,322. 7734	650	1,433. 0045	700	1,543. 2356
1	1,104. 5158	1	1,214. 7469	1	1,324. 9780	1	1,435. 2091	1	1,545. 4403
2	1,106. 7204	2	1,216. 9515	2	1,327. 1826	2	1,437. 4138	2	1,547. 6449
3	1,108. 9250	3	1,219. 1562	3	1,329. 3873	3	1,439. 6184	3	1,549. 8495
4	1,111. 1297	4	1,221. 3608	4	1,331. 5919	4	1,441. 8230	4	1,552. 0541
5	1,113. 3343	5	1,223. 5654	5	1,333. 7965	5	1,444. 0276	5	1,554. 2588
6	1,115. 5389	6	1,225. 7700	6	1,336. 0011	6	1,446. 2323	6	1,556. 4634
7	1,117. 7435	7	1,227. 9746	7	1,338. 2058	7	1,448. 4369	7	1,558. 6680
8	1,119. 9481	8	1,230. 1793	8	1,340. 4104	8	1,450. 6415	8	1,560. 8726
9	1,122. 1528	9	1,232. 3839	9	1,342. 6150	9	1,452. 8461	9	1,563. 0772
510	1,124. 3574	560	1,234. 5885	610	1,344. 8196	660	1,455. 0507	710	1,565. 2819
1	1,126. 5620	1	1,236. 7931	1	1,347. 0243	1	1,457. 2554	1	1,567. 4865
2	1,128. 7666	2	1,238. 9978	2	1,349. 2289	2	1,459. 4600	2	1,569. 6911
3	1,130. 9713	3	1,241. 2024	3	1,351. 4335	3	1,461. 6646	3	1,571. 8957
4	1,133. 1759	4	1,243. 4070	4	1,353. 6381	4	1,463. 8692	4	1,574. 1004
5	1,135. 3805	5	1,245. 6116	5	1,355. 8427	5	1,466. 0739	5	1,576. 3050
6	1,137. 5851	6	1,247. 8162	6	1,358. 0474	6	1,468. 2785	6	1,578. 5096
7	1,139. 7898	7	1,250. 0209	7	1,360. 2520	7	1,470. 4831	7	1,580. 7142
8	1,141. 9944	8	1,252. 2255	8	1,362. 4566	8	1,472. 6877	8	1,582. 9188
9	1,144. 1990	9	1,254. 4301	9	1,364. 6612	9	1,474. 8923	9	1,585. 1235
520	1,146. 4036	570	1,256. 6347	620	1,366. 8659	670	1,477. 0970	720	1,587. 3281
1	1,148. 6082	1	1,258. 8394	1	1,369. 0705	1	1,479. 3016	1	1,589. 5327
2	1,150. 8129	2	1,261. 0440	2	1,371. 2751	2	1,481. 5062	2	1,591. 7373
3	1,153. 0175	3	1,263. 2486	3	1,373. 4797	3	1,483. 7108	3	1,593. 9420
4	1,155. 2221	4	1,265. 4532	4	1,375. 6843	4	1,485. 9155	4	1,596. 1466
5	1,157. 4267	5	1,267. 6578	5	1,377. 8890	5	1,488. 1201	5	1,598. 3512
6	1,159. 6314	6	1,269. 8625	6	1,380. 0936	6	1,490. 3247	6	1,600. 5558
7	1,161. 8360	7	1,272. 0671	7	1,382. 2982	7	1,492. 5293	7	1,602. 7604
8	1,164. 0406	8	1,274. 2717	8	1,384. 5028	8	1,494. 7339	8	1,604. 9651
9	1,166. 2452	9	1,276. 4763	9	1,386. 7075	9	1,496. 9386	9	1,607. 1697
530	1,168. 4498	580	1,278. 6810	630	1,388. 9121	680	1,499. 1432	730	1,609. 3743
1	1,170. 6545	1	1,280. 8856	1	1,391. 1167	1	1,501. 3478	1	1,611. 5789
2	1,172. 8591	2	1,283. 0902	2	1,393. 3213	2	1,503. 5524	2	1,613. 7836
3	1,175. 0637	3	1,285. 2948	3	1,395. 5259	3	1,505. 7571	3	1,615. 9882
4	1,177. 2683	4	1,287. 4994	4	1,397. 7306	4	1,507. 9617	4	1,618. 1928
5	1,179. 4730	5	1,289. 7041	5	1,399. 9352	5	1,510. 1663	5	1,620. 3974
6	1,181. 6776	6	1,291. 9087	6	1,402. 1398	6	1,512. 3709	6	1,622. 6020
7	1,183. 8822	7	1,294. 1133	7	1,404. 3444	7	1,514. 5755	7	1,624. 8067
8	1,186. 0868	8	1,296. 3179	8	1,406. 5491	8	1,516. 7802	8	1,627. 0113
9	1,188. 2914	9	1,298. 5226	9	1,408. 7537	9	1,518. 9848	9	1,629. 2159
540	1,190. 4961	590	1,300. 7272	640	1,410. 9583	690	1,521. 1894	740	1,631. 4205
1	1,192. 7007	1	1,302. 9318	1	1,413. 1629	1	1,523. 3940	1	1,633. 6252
2	1,194. 9053	2	1,305. 1364	2	1,415. 3675	2	1,525. 5987	2	1,635. 8298
3	1,197. 1099	3	1,307. 3410	3	1,417. 5722	3	1,527. 8033	3	1,638. 0344
4	1,199. 3146	4	1,309. 5457	4	1,419. 7768	4	1,530. 0079	4	1,640. 2390
5	1,201. 5192	5	1,311. 7503	5	1,421. 9814	5	1,532. 2125	5	1,642. 4436
6	1,203. 7238	6	1,313. 9549	6	1,424. 1860	6	1,534. 4171	6	1,644. 6483
7	1,205. 9284	7	1,316. 1595	7	1,426. 3907	7	1,536. 6218	7	1,646. 8529
8	1,208. 1330	8	1,318. 3642	8	1,428. 5953	8	1,538. 8264	8	1,649. 0575
9	1,210. 3377	9	1,320. 5688	9	1,430. 7999	9	1,541. 0310	9	1,651. 2621

MASS—KILOGRAMS TO AVOIRDUPOIS POUNDS

[Reduction factor: 1 kilogram = 2.204622341 avoirdupois pounds]

Kilos	Pounds	Kilos	Pounds	Kilos	Pounds	Kilos	Pounds	Kilos	Pounds
750	1,653.4668	800	1,763.6979	850	1,873.9290	900	1,984.1601	950	2,094.3912
1	1,655.6714	1	1,765.9025	1	1,876.1336	1	1,986.3647	1	2,096.5958
2	1,657.8760	2	1,768.1071	2	1,878.3382	2	1,988.5694	2	2,098.8005
3	1,660.0806	3	1,770.3117	3	1,880.5429	3	1,990.7740	3	2,101.0051
4	1,662.2852	4	1,772.5164	4	1,882.7475	4	1,992.9786	4	2,103.2097
5	1,664.4899	5	1,774.7210	5	1,884.9521	5	1,995.1832	5	2,105.4143
6	1,666.6945	6	1,776.9256	6	1,887.1567	6	1,997.3878	6	2,107.6190
7	1,668.8991	7	1,779.1302	7	1,889.3613	7	1,999.5925	7	2,109.8236
8	1,671.1037	8	1,781.3349	8	1,891.5660	8	2,001.7971	8	2,112.0282
9	1,673.3084	9	1,783.5395	9	1,893.7706	9	2,004.0017	9	2,114.2328
760	1,675.5130	810	1,785.7441	860	1,895.9752	910	2,006.2063	960	2,116.4374
1	1,677.7176	1	1,787.9487	1	1,898.1798	1	2,008.4110	1	2,118.6421
2	1,679.9222	2	1,790.1533	2	1,900.3845	2	2,010.6156	2	2,120.8467
3	1,682.1268	3	1,792.3580	3	1,902.5891	3	2,012.8202	3	2,123.0513
4	1,684.3315	4	1,794.5626	4	1,904.7937	4	2,015.0248	4	2,125.2559
5	1,686.5361	5	1,796.7672	5	1,906.9983	5	2,017.2294	5	2,127.4606
6	1,688.7407	6	1,798.9718	6	1,909.2029	6	2,019.4341	6	2,129.6652
7	1,690.9453	7	1,801.1765	7	1,911.4076	7	2,021.6387	7	2,131.8698
8	1,693.1500	8	1,803.3811	8	1,913.6122	8	2,023.8433	8	2,134.0744
9	1,695.3546	9	1,805.5857	9	1,915.8168	9	2,026.0479	9	2,136.2790
770	1,697.5592	820	1,807.7903	870	1,918.0214	920	2,028.2526	970	2,138.4837
1	1,699.7638	1	1,809.9949	1	1,920.2261	1	2,030.4572	1	2,140.6883
2	1,701.9684	2	1,812.1996	2	1,922.4307	2	2,032.6618	2	2,142.8929
3	1,704.1731	3	1,814.4042	3	1,924.6353	3	2,034.8664	3	2,145.0975
4	1,706.3777	4	1,816.6088	4	1,926.8399	4	2,037.0710	4	2,147.3022
5	1,708.5823	5	1,818.8134	5	1,929.0445	5	2,039.2757	5	2,149.5068
6	1,710.7869	6	1,821.0181	6	1,931.2492	6	2,041.4803	6	2,151.7114
7	1,712.9916	7	1,823.2227	7	1,933.4538	7	2,043.6849	7	2,153.9160
8	1,715.1962	8	1,825.4273	8	1,935.6584	8	2,045.8895	8	2,156.1206
9	1,717.4008	9	1,827.6319	9	1,937.8630	9	2,048.0942	9	2,158.3253
780	1,719.6054	830	1,829.8365	880	1,940.0677	930	2,050.2988	980	2,160.5299
1	1,721.8100	1	1,832.0412	1	1,942.2723	1	2,052.5034	1	2,162.7345
2	1,724.0147	2	1,834.2458	2	1,944.4769	2	2,054.7080	2	2,164.9391
3	1,726.2193	3	1,836.4504	3	1,946.6815	3	2,056.9126	3	2,167.1438
4	1,728.4239	4	1,838.6550	4	1,948.8861	4	2,059.1173	4	2,169.3484
5	1,730.6285	5	1,840.8597	5	1,951.0908	5	2,061.3219	5	2,171.5530
6	1,732.8332	6	1,843.0643	6	1,953.2954	6	2,063.5265	6	2,173.7576
7	1,735.0378	7	1,845.2689	7	1,955.5000	7	2,065.7311	7	2,175.9623
8	1,737.2424	8	1,847.4735	8	1,957.7046	8	2,067.9358	8	2,178.1669
9	1,739.4470	9	1,849.6781	9	1,959.9093	9	2,070.1404	9	2,180.3715
790	1,741.6516	840	1,851.8828	890	1,962.1139	940	2,072.3450	990	2,182.5761
1	1,743.8563	1	1,854.0874	1	1,964.3185	1	2,074.5496	1	2,184.7807
2	1,746.0609	2	1,856.2920	2	1,966.5231	2	2,076.7542	2	2,186.9854
3	1,748.2655	3	1,858.4966	3	1,968.7278	3	2,078.9589	3	2,189.1900
4	1,750.4701	4	1,860.7013	4	1,970.9324	4	2,081.1635	4	2,191.3946
5	1,752.6748	5	1,862.9059	5	1,973.1370	5	2,083.3681	5	2,193.5992
6	1,754.8794	6	1,865.1105	6	1,975.3416	6	2,085.5727	6	2,195.8039
7	1,757.0840	7	1,867.3151	7	1,977.5462	7	2,087.7774	7	2,198.0085
8	1,759.2886	8	1,869.5197	8	1,979.7509	8	2,089.9820	8	2,200.2131
9	1,761.4933	9	1,871.7244	9	1,981.9555	9	2,092.1866	9	2,202.4177

MASS—AVOIRDUPOIS POUNDS TO KILOGRAMS

[Reduction factor: 1 avoirdupois pound = 0.4535924277 kilogram]

Pounds	Kilos	Pounds	Kilos	Pounds	Kilos	Pounds	Kilos	Pounds	Kilos
0		50	22.67962	100	45.35924	150	68.03886	200	90.71849
1	0.45359	1	23.13321	1	45.81284	1	68.49246	1	91.17208
2	.90718	2	23.58681	2	46.26643	2	68.94605	2	91.62567
3	1.36078	3	24.04040	3	46.72002	3	69.39964	3	92.07926
4	1.81437	4	24.49399	4	47.17361	4	69.85323	4	92.53286
5	2.26796	5	24.94758	5	47.62720	5	70.30683	5	92.98645
6	2.72155	6	25.40118	6	48.08080	6	70.76042	6	93.44004
7	3.17515	7	25.85477	7	48.53439	7	71.21401	7	93.89363
8	3.62874	8	26.30836	8	48.98798	8	71.66760	8	94.34722
9	4.08233	9	26.76195	9	49.44157	9	72.12120	9	94.80082
10	4.53592	60	27.21555	110	49.89517	160	72.57479	210	95.25441
1	4.98952	1	27.66914	1	50.34876	1	73.02838	1	95.70800
2	5.44311	2	28.12273	2	50.80235	2	73.48197	2	96.16159
3	5.89670	3	28.57632	3	51.25594	3	73.93557	3	96.61519
4	6.35029	4	29.02992	4	51.70954	4	74.38916	4	97.06878
5	6.80389	5	29.48351	5	52.16313	5	74.84275	5	97.52237
6	7.25748	6	29.93710	6	52.61672	6	75.29634	6	97.97596
7	7.71107	7	30.39069	7	53.07031	7	75.74994	7	98.42956
8	8.16466	8	30.84429	8	53.52391	8	76.20353	8	98.88315
9	8.61826	9	31.29788	9	53.97750	9	76.65712	9	99.33674
20	9.07185	70	31.75147	120	54.43109	170	77.11071	220	99.79033
1	9.52544	1	32.20506	1	54.88468	1	77.56431	1	100.24393
2	9.97903	2	32.65865	2	55.33828	2	78.01790	2	100.69752
3	10.43263	3	33.11225	3	55.79187	3	78.47149	3	101.15111
4	10.88622	4	33.56584	4	56.24546	4	78.92509	4	101.60470
5	11.33981	5	34.01943	5	56.69905	5	79.37867	5	102.05830
6	11.79340	6	34.47302	6	57.15265	6	79.83227	6	102.51189
7	12.24700	7	34.92662	7	57.60624	7	80.28586	7	102.96548
8	12.70059	8	35.38021	8	58.05983	8	80.73945	8	103.41907
9	13.15418	9	35.83380	9	58.51342	9	81.19304	9	103.87267
30	13.60777	80	36.28739	130	58.96702	180	81.64664	230	104.32626
1	14.06137	1	36.74099	1	59.42061	1	82.10023	1	104.77985
2	14.51496	2	37.19458	2	59.87420	2	82.55382	2	105.23344
3	14.96855	3	37.64817	3	60.32779	3	83.00741	3	105.68704
4	15.42214	4	38.10176	4	60.78139	4	83.46101	4	106.14063
5	15.87573	5	38.55536	5	61.23498	5	83.91460	5	106.59422
6	16.32933	6	39.00895	6	61.68857	6	84.36819	6	107.04781
7	16.78292	7	39.46254	7	62.14216	7	84.82178	7	107.50141
8	17.23651	8	39.91613	8	62.59576	8	85.27538	8	107.95500
9	17.69010	9	40.36973	9	63.04935	9	85.72897	9	108.40859
40	18.14370	90	40.82332	140	63.50294	190	86.18256	240	108.86218
1	18.59729	1	41.27691	1	63.95653	1	86.63615	1	109.31578
2	19.05088	2	41.73050	2	64.41012	2	87.08975	2	109.76937
3	19.50447	3	42.18410	3	64.86372	3	87.54334	3	110.22296
4	19.95807	4	42.63769	4	65.31731	4	87.99693	4	110.67655
5	20.41166	5	43.09128	5	65.77090	5	88.45052	5	111.13014
6	20.86525	6	43.54487	6	66.22449	6	88.90412	6	111.58374
7	21.31884	7	43.99847	7	66.67809	7	89.35771	7	112.03733
8	21.77244	8	44.45206	8	67.13168	8	89.81130	8	112.49092
9	22.22603	9	44.90565	9	67.58527	9	90.26489	9	112.94451

MASS—AVOIRDUPOIS POUNDS TO KILOGRAMS

[Reduction factor: 1 avoirdupois pound = 0.4535924277 kilogram]

Pounds	Kilos	Pounds	Kilos	Pounds	Kilos	Pounds	Kilos	Pounds	Kilos
250	113. 39811	**300**	136. 07773	**350**	158. 75735	**400**	181. 43697	**450**	204. 11659
1	113. 85170	1	136. 53132	1	159. 21094	1	181. 89056	1	204. 57018
2	114. 30529	2	136. 98491	2	159. 66453	2	182. 34416	2	205. 02378
3	114. 75888	3	137. 43851	3	160. 11813	3	182. 79775	3	205. 47737
4	115. 21248	4	137. 89210	4	160. 57172	4	183. 25134	4	205. 93096
5	115. 66607	5	138. 34569	5	161. 02531	5	183. 70493	5	206. 38455
6	116. 11966	6	138. 79928	6	161. 47890	6	184. 15853	6	206. 83815
7	116. 57325	7	139. 25288	7	161. 93250	7	184. 61212	7	207. 29174
8	117. 02685	8	139. 70647	8	162. 38609	8	185. 06571	8	207. 74533
9	117. 48044	9	140. 16006	9	162. 83968	9	185. 51930	9	208. 19892
260	117. 93403	**310**	140. 61365	**360**	163. 29327	**410**	185. 97290	**460**	208. 65252
1	118. 38762	1	141. 06725	1	163. 74687	1	186. 42649	1	209. 10611
2	118. 84122	2	141. 52084	2	164. 20046	2	186. 88008	2	209. 55970
3	119. 29481	3	141. 97443	3	164. 65405	3	187. 33367	3	210. 01329
4	119. 74840	4	142. 42802	4	165. 10764	4	187. 78727	4	210. 46689
5	120. 20199	5	142. 88161	5	165. 56124	5	188. 24086	5	210. 92048
6	120. 65559	6	143. 33521	6	166. 01483	6	188. 69445	6	211. 37407
7	121. 10918	7	143. 78880	7	166. 46842	7	189. 14804	7	211. 82766
8	121. 56277	8	144. 24239	8	166. 92201	8	189. 60163	8	212. 28126
9	122. 01636	9	144. 69598	9	167. 37561	9	190. 05523	9	212. 73485
270	122. 46996	**320**	145. 14958	**370**	167. 82920	**420**	190. 50882	**470**	213. 18844
1	122. 92355	1	145. 60317	1	168. 28279	1	190. 96241	1	213. 64203
2	123. 37714	2	146. 05676	2	168. 73638	2	191. 41600	2	214. 09563
3	123. 83073	3	146. 51035	3	169. 18998	3	191. 86960	3	214. 54922
4	124. 28433	4	146. 96395	4	169. 64357	4	192. 32319	4	215. 00281
5	124. 73792	5	147. 41754	5	170. 09716	5	192. 77678	5	215. 45640
6	125. 19151	6	147. 87113	6	170. 55075	6	193. 23037	6	215. 91000
7	125. 64510	7	148. 32472	7	171. 00435	7	193. 68397	7	216. 36359
8	126. 09869	8	148. 77832	8	171. 45794	8	194. 13756	8	216. 81718
9	126. 55229	9	149. 23191	9	171. 91153	9	194. 59115	9	217. 27077
280	127. 00588	**330**	149. 68550	**380**	172. 36512	**430**	195. 04474	**480**	217. 72437
1	127. 45947	1	150. 13909	1	172. 81871	1	195. 49834	1	218. 17796
2	127. 91306	2	150. 59269	2	173. 27231	2	195. 95193	2	218. 63155
3	128. 36666	3	151. 04628	3	173. 72590	3	196. 40552	3	219. 08514
4	128. 82025	4	151. 49987	4	174. 17949	4	196. 85911	4	219. 53874
5	129. 27384	5	151. 95346	5	174. 63308	5	197. 31271	5	219. 99233
6	129. 72743	6	152. 40706	6	175. 08668	6	197. 76630	6	220. 44592
7	130. 18103	7	152. 86065	7	175. 54027	7	198. 21989	7	220. 89951
8	130. 63462	8	153. 31424	8	175. 99386	8	198. 67348	8	221. 35310
9	131. 08821	9	153. 76783	9	176. 44745	9	199. 12708	9	221. 80670
290	131. 54180	**340**	154. 22143	**390**	176. 90105	**440**	199. 58067	**490**	222. 26029
1	131. 99540	1	154. 67502	1	177. 35464	1	200. 03426	1	222. 71388
2	132. 44899	2	155. 12861	2	177. 80823	2	200. 48785	2	223. 16747
3	132. 90258	3	155. 58220	3	178. 26182	3	200. 94145	3	223. 62107
4	133. 35617	4	156. 03580	4	178. 71542	4	201. 39504	4	224. 07466
5	133. 80977	5	156. 48939	5	179. 16901	5	201. 84863	5	224. 52825
6	134. 26336	6	156. 94298	6	179. 62260	6	202. 30222	6	224. 98184
7	134. 71695	7	157. 39657	7	180. 07619	7	202. 75582	7	225. 43544
8	135. 17054	8	157. 85016	8	180. 52979	8	203. 20941	8	225. 88903
9	135. 62414	9	158. 30376	9	180. 98338	9	203. 66300	9	226. 34262

MASS—AVOIRDUPOIS POUNDS TO KILOGRAMS

[Reduction factor: 1 avoirdupois pound = 0.4535924277 kilogram]

Pounds	Kilos	Pounds	Kilos	Pounds	Kilos	Pounds	Kilos	Pounds	Kilos
500	226.79621	550	249.47584	600	272.15546	650	294.83508	700	317.51470
1	227.24981	1	249.92943	1	272.60905	1	295.28867	1	317.96829
2	227.70340	2	250.38302	2	273.06264	2	295.74226	2	318.42188
3	228.15699	3	250.83661	3	273.51623	3	296.19586	3	318.87548
4	228.61058	4	251.29020	4	273.96983	4	296.64945	4	319.32907
5	229.06418	5	251.74380	5	274.42342	5	297.10304	5	319.78266
6	229.51777	6	252.19739	6	274.87701	6	297.55663	6	320.23625
7	229.97136	7	252.65098	7	275.33060	7	298.01022	7	320.68985
8	230.42495	8	253.10457	8	275.78420	8	298.46382	8	321.14344
9	230.87855	9	253.55817	9	276.23779	9	298.91741	9	321.59703
510	231.33214	560	254.01176	610	276.69138	660	299.37100	710	322.05062
1	231.78573	1	254.46535	1	277.14497	1	299.82459	1	322.50422
2	232.23932	2	254.91894	2	277.59857	2	300.27819	2	322.95781
3	232.69292	3	255.37254	3	278.05216	3	300.73178	3	323.41140
4	233.14651	4	255.82613	4	278.50575	4	301.18537	4	323.86499
5	233.60010	5	256.27972	5	278.95934	5	301.63896	5	324.31859
6	234.05369	6	256.73331	6	279.41294	6	302.09256	6	324.77218
7	234.50729	7	257.18691	7	279.86653	7	302.54615	7	325.22577
8	234.96088	8	257.64050	8	280.32012	8	302.99974	8	325.67936
9	235.41447	9	258.09409	9	280.77371	9	303.45333	9	326.13296
520	235.86806	570	258.54768	620	281.22731	670	303.90693	720	326.58655
1	236.32165	1	259.00128	1	281.68090	1	304.36052	1	327.04014
2	236.77525	2	259.45487	2	282.13449	2	304.81411	2	327.49373
3	237.22884	3	259.90846	3	282.58808	3	305.26770	3	327.94733
4	237.68243	4	260.36205	4	283.04167	4	305.72130	4	328.40092
5	238.13602	5	260.81565	5	283.49527	5	306.17489	5	328.85451
6	238.58962	6	261.26924	6	283.94886	6	306.62848	6	329.30810
7	239.04321	7	261.72283	7	284.40245	7	307.08207	7	329.76169
8	239.49680	8	262.17642	8	284.85604	8	307.53567	8	330.21529
9	239.95039	9	262.63002	9	285.30964	9	307.98926	9	330.66888
530	240.40399	580	263.08361	630	285.76323	680	308.44285	730	331.12247
1	240.85758	1	263.53720	1	286.21682	1	308.89644	1	331.57606
2	241.31117	2	263.99079	2	286.67041	2	309.35004	2	332.02966
3	241.76476	3	264.44439	3	287.12401	3	309.80363	3	332.48325
4	242.21836	4	264.89798	4	287.57760	4	310.25722	4	332.93684
5	242.67195	5	265.35157	5	288.03119	5	310.71081	5	333.39043
6	243.12554	6	265.80516	6	288.48478	6	311.16441	6	333.84403
7	243.57913	7	266.23876	7	288.93838	7	311.61000	7	334.29762
8	244.03273	8	266.71235	8	289.39197	8	312.07159	8	334.75121
9	244.48632	9	267.16594	9	289.84556	9	312.52518	9	335.20480
540	244.93991	590	267.61953	640	290.29915	690	312.97878	740	335.65840
1	245.39350	1	268.07312	1	290.75275	1	313.43237	1	336.11199
2	245.84710	2	268.52672	2	291.20634	2	313.88596	2	336.56558
3	246.30069	3	268.98031	3	291.65993	3	314.33955	3	337.01917
4	246.75428	4	269.43390	4	292.11352	4	314.79314	4	337.47277
5	247.20787	5	269.88749	5	292.56712	5	315.24674	5	337.92636
6	247.66147	6	270.34109	6	293.02071	6	315.70033	6	338.37995
7	248.11506	7	270.79468	7	293.47430	7	316.15392	7	338.83354
8	248.56865	8	271.24827	8	293.92789	8	316.60751	8	339.28714
9	249.02224	9	271.70186	9	294.38149	9	317.06111	9	339.74073

MASS—AVOIRDUPOIS POUNDS TO KILOGRAMS

[Reduction factor: 1 avoirdupois pound = 0.4535924277 kilogram]

Pounds	Kilos	Pounds	Kilos	Pounds	Kilos	Pounds	Kilos	Pounds	Kilos
750	340.19432	800	362.87394	850	385.55356	900	408.23318	950	430.91281
1	340.64791	1	363.32753	1	386.00716	1	408.68678	1	431.36640
2	341.10151	2	363.78113	2	386.46075	2	409.14037	2	431.81999
3	341.55510	3	364.23472	3	386.91434	3	409.59396	3	432.27358
4	342.00869	4	364.68831	4	387.36793	4	410.04755	4	432.72718
5	342.46228	5	365.14190	5	387.82153	5	410.50115	5	433.18077
6	342.91588	6	365.59550	6	388.27512	6	410.95474	6	433.63436
7	343.36947	7	366.04909	7	388.72871	7	411.40833	7	434.08795
8	343.82306	8	366.50268	8	389.18230	8	411.86192	8	434.54155
9	344.27665	9	366.95627	9	389.63590	9	412.31552	9	434.99514
760	344.73025	810	367.40987	860	390.08949	910	412.76911	960	435.44873
1	345.18384	1	367.86346	1	390.54308	1	413.22270	1	435.90232
2	345.63743	2	368.31705	2	390.99667	2	413.67629	2	436.35592
3	346.09102	3	368.77064	3	391.45027	3	414.12989	3	436.80951
4	346.54461	4	369.22424	4	391.90386	4	414.58348	4	437.26310
5	346.99821	5	369.67783	5	392.35745	5	415.03707	5	437.71669
6	347.45180	6	570.13142	6	392.81104	6	415.49066	6	438.17029
7	347.90539	7	370.58501	7	393.26463	7	415.94426	7	438.62388
8	348.35898	8	371.03861	8	393.71823	8	416.39785	8	439.07747
9	348.81258	9	371.49220	9	394.17182	9	416.85144	9	439.53106
770	349.26617	820	371.94579	870	394.62541	920	417.30503	970	439.98465
1	349.71976	1	372.39938	1	395.07900	1	417.75863	1	440.43825
2	350.17335	2	372.85298	2	395.53260	2	418.21222	2	440.89184
3	350.62695	3	373.30657	3	395.98619	3	418.66581	3	441.34543
4	351.08054	4	373.76016	4	396.43978	4	419.11940	4	441.79902
5	351.53413	5	374.21375	5	396.89337	5	419.57300	5	442.25262
6	351.98772	6	374.66735	6	397.34697	6	420.02659	6	442.70621
7	352.44132	7	375.12094	7	397.80056	7	420.48018	7	443.15980
8	352.89491	8	375.57453	8	398.25415	8	420.93377	8	443.61339
9	353.34850	9	376.02812	9	398.70774	9	421.38737	9	444.06699
780	353.80209	830	376.48171	880	399.16134	930	421.84096	980	444.52058
1	354.25569	1	376.93531	1	399.61493	1	422.29455	1	444.97417
2	354.70928	2	377.38890	2	400.06852	2	422.74814	2	445.42776
3	355.16287	3	377.84249	3	400.52211	3	423.20174	3	445.88136
4	355.61646	4	378.29608	4	400.97571	4	423.65533	4	446.33495
5	356.07006	5	378.74968	5	401.42930	5	424.10892	5	446.78854
6	356.52365	6	379.20327	6	401.88289	6	424.56251	6	447.24213
7	356.97724	7	379.65686	7	402.33648	7	425.01610	7	447.69573
8	357.43083	8	380.11045	8	402.79008	8	425.46970	8	448.14932
9	357.88443	9	380.56405	9	403.24367	9	425.92329	9	448.60291
790	358.33802	840	381.01764	890	403.69726	940	426.37688	990	449.05650
1	358.79161	1	381.47123	1	404.15085	1	426.83047	1	449.51010
2	359.24520	2	381.92482	2	404.60445	2	427.28407	2	449.96369
3	359.69880	3	382.37842	3	405.05804	3	427.73766	3	450.41728
4	360.15239	4	382.83201	4	405.51163	4	428.19125	4	450.87087
5	360.60598	5	383.28560	5	405.96522	5	428.64484	5	451.32447
6	361.05957	6	383.73919	6	406.41882	6	429.09844	6	451.77806
7	361.51316	7	384.19279	7	406.87241	7	429.55203	7	452.23165
8	361.96676	8	384.64638	8	407.32600	8	430.00562	8	452.68524
9	362.42035	9	385.09997	9	407.77959	9	430.45921	9	453.13884

METRIC AND ENGLISH EQUIVALENTS OF DISTANCE IN TRACK AND FIELD EVENTS

Metric distances for track and field events to be run in athletic meets held under the jurisdiction of the Amateur Athletic Union were officially adopted by that body on November 22, 1932. The following tables have been included in this book to make it possible for those not familiar with the Metric system to know the various distances expressed in metres.

In Table 1 are given the equivalents of Metric and English distances for principal indoor and outdoor track events.

In Table 2 are given the Metric equivalents for distances in feet, inches and binary fractions of an inch.

The metric equivalent of any distance may be conveniently found, to the nearest ⅛ inch, by breaking the distance down into convenient parts, obtaining the equivalent of each part and then adding them together to get the total equivalent.

For example the metric equivalent of 251 feet, 9½ inches is found as follows:

$$
\begin{array}{rcl}
200 \text{ feet} & = & 60.960 \text{ metres} \\
50 \text{ feet} & = & 15.240 \text{ metres} \\
1 \text{ foot} & = & .0305 \text{ metre} \\
9 \text{ inches} & = & .229 \text{ metre} \\
\tfrac{1}{2} \text{ inch} & = & .013 \text{ metre} \\
\hline
251 \text{ feet}, 9\tfrac{1}{2} \text{ inches} & = & 76.4725 \text{ metres}
\end{array}
$$

DISTANCE EQUIVALENTS

Basis $\begin{cases} 1 \text{ metre} = 39.37 \text{ inches} = 3.280\ 8 \text{ feet} = 1.093\ 6 \text{ yards} \\ 1 \text{ kilometre} = 1\ 000 \text{ metres} = 0.621\ 370 \text{ mile} \end{cases}$

TABLE 1.—*Track events*

Yards : Metres		Metres : Yards	
40 =	36. 58	50 =	54. 68
50 =	45. 72	60 =	65. 62
60 =	54. 86	65 =	71. 08
70 =	64. 01	80 =	87. 49
75 =	68. 58	100 =	109. 36
100 =	91. 44	110 =	120. 30
110 =	100. 58	200 =	218. 72
120 =	109. 73	300 =	328. 08
220 =	201. 17	400 =	437. 44
300 =	274. 32	500 =	546. 81
440 =	402. 34 = ¼ mi	600 =	656. 16
600 =	548. 64	800 =	874. 89
880 =	804. 67 = ½ mi	1 000 =	1 093. 61
1 000 =	914. 40	1 500 =	1 640. 42
1 320 =	1 207. 01 = ¾ mi	1 600 =	1 749. 78

Miles : Metres		Metres : Miles		Yards and inches		Miles (approx.)
1 =	1 609. 3	2 000 =	1	427	8	1. 24
2 =	3 218. 7	2 400 =	1	864	24	1. 49
3 =	4 828. 0	3 000 =	1	1 520	30	1. 86
4 =	6 437. 4	3 200 =	1	1 739	20	1. 99
5 =	8 046. 7					
		5 000 =	3	188	2	3. 11
6 =	9 656. 1	6 000 =	3	1 281	24	3. 73
7 =	11 265. 4	10 000 =	6	376	4	6. 21
8 =	12 874. 8	15 000 =	9	564	6	9. 32
9 =	14 484. 1					
		20 000 =	12	752	8	12. 43
10 =	16 093. 5	25 000 =	15	940	10	15. 53
15 =	24 140. 2	30 000 =	18	1 128	12	18. 64
20 =	32 186. 9	50 000 =	31	120	20	31. 07
25 =	40 233. 7					

TABLE 2.—*Field events*

Feet : Metres		Inches : Metres	
1	= 0.305	1 =	0.025
2	= .610	2 =	.051
3	= .914	3 =	.076
4	= 1.219	4 =	.102
5	= 1.524		
		5 =	.127
6	= 1.829	6 =	.152
7	= 2.134	7 =	.178
8	= 2.438	8 =	.203
9	= 2.743		
		9 =	.229
10	= 3.048	10 =	.254
20	= 6.096	11 =	.279
30	= 9.144	12 =	.305
40	= 12.192		
50	= 15.240	Fractions of an inch : Metre	
60	= 18.288		
70	= 21.336	$\frac{1}{8}$ =	0.003
80	= 24.384	$\frac{1}{4}$ =	.006
90	= 27.432	$\frac{3}{8}$ =	.010
		$\frac{1}{2}$ =	.013
100	= 30.480		
200	= 60.960	$\frac{5}{8}$ =	.016
300	= 91.440	$\frac{3}{4}$ =	.019
400	= 121.920	$\frac{7}{8}$ =	.022
500	= 152.400	1 =	.025
600	= 182.880		
700	= 213.360		
800	= 243.840		
900	= 274.321		

The precision of measurement of distance and of time, as ordinarily carried out in track and field events, received consideration in determining the number of decimal places to be carried out in these tables.

Distances in field events are customarily measured in feet and inches to the nearest eighth of an inch. This is about 3 millimetres or 0.0003 metres. In order to convert these measured distances from feet and inches to metres with maximum accuracy, the metric equivalents are given to the nearest 0.001 metre.

The same consideration has been given to the accuracy of measurement of both distance and time as carried out with track events. It should be noted that when time is taken with a stop watch it is usually given to the fifth or tenth of a second. When taken with electrical timing devices it may be given to the hundredth of a second.

In dashes, where 1 second represents a distance of approximately 10 yards or 10 metres, $\frac{1}{10}$ second represents about 1 yard or 1 metre, and $\frac{1}{100}$ second represents a distance of $\frac{1}{10}$ yard or $\frac{1}{10}$ metre. There is no need at present, therefore, to give metric equivalents of distances more precision than the nearest $\frac{1}{10}$ metre, even when the most precise timing methods are used. For distances of less than a mile, however, they have been given to the nearest $\frac{1}{100}$ metre in order to allow for possible future improvement in timing devices.

INDEX

ANSWERS TO PRACTICE EXAMPLES

Page 20 - Exercise 1

Group 1: (a) 950 mm   (b) 235 mm   (c) 80 mm
          (d) 95 mm   (e) 120 mm
Group 2: (a) 82.5 cm   (b) 400 cm   (c) 62.0 cm
          (d) 90 cm   (e) 32.5 cm
Group 3: (a) 300 dm   (b) 6.25 dm   (c) 80 dm
          (d) 184.5 dm   (e) 30 dm
Group 4: (a) 800 cm   (b) 75 cm   (d) 760 cm
          (d) 1500 cm   (e) 6 500 cm

Exercise 2

Group 5: (a) 9 000 mm   (b) 15 500 mm   (c) 25 000 mm
          (d) 6 330 mm   (e) 12 125 mm
Group 6: (a) 200 m   (b) 40 m   (c) 750 m   (d) 84 m
          (e) 46.4 m
Group 7: (a) 100 dam   (b) 92.5 dam   (c) 450 dam
          (d) 26.25 dam   (e) 170 dam
Group 8: (a) 50 hm   (b) 160 hm   (c) 630 hm
          (d) 2.5 hm   (e) 86.3 hm

Page 21 - Exercise 3

Group 9: (a) 600 m   (b) 1 520 m
          (c) 2 400 m (d) 37.5 m   (e) 8 900 m
Group 10: (a) 5 000 m   (b) 53 000 m   (c) 121 000 m
           (d) 37 500 m   (e) 125 m
Group 11: (a) 4.5 cm   (b) 2.17 cm   (c) 1.6 cm
           (d) 87.5 cm   (e) 3.25 cm
Group 12: (a) 3 dm   (b) 5.5 dm   (c) 0.8 dm
           (d) 2.53 dm   (e) 0.06 dm

Exercise 4

Group 13: (a) 12.5 m   (b) 1.5 m   (c) 0.4 m
           (d) 0.725 m   (e) 1.25 m
Group 14: (a) 3.5 m   (b) 2.22 m   (c) 16.24 m
           (d) 6.355 m   (e) 0.33 m
Group 15: (a) 2.5 m   (b) 1.25 m   (c) 10.375 m
           (d) 0.2395 m   (e) 0.055 5 m
Group 16: (a) 7.5 dam   (b) 18.55 dam   (c) 2.95 dam
           (d) 0.8 dam   (e) 0.63 dam

ANSWERS TO PRACTICE EXAMPLES (continued)

Page 22 - Exercise 5

Group 17: (a) 17.5 hm   (b) 3.8 hm   (c) 0.6 hm
          (d) 5.75 hm   (e) 0.09 hm
Group 18: (a) 6.5 km   (b) 0.7 km   (c) 1.625 km
          (d) 22.55 km   (e) 2.9 km
Group 19: (a) 5.5 hm   (b) 14.87 hm   (c) 3.4 hm
          (d) 21.25 hm   (e) 0.44 hm
Group 20: (a) 0.35 km   (b) 7.85 km   (c) 0.3092 km
          (d) 1.728 km   (e) 2 km

Exercise 6

Group 21: (a) 300 000 cm   (b) 62 500 cm
          (c) 1 800 000 cm   (d) 725 000 cm
          (e) 162 500 cm
Group 22: (a) 6.5 km   (b) 12.5 km   (c) 2.535 km
          (d) 0.45 km   (e) 1.784 km
Group 23: (a) 15 000 000 mm   (b) 4 000 000 mm
          (c) 2 250 000 mm   (d) 500 000 mm
          (e) 1 750 000 mm
Group 24: (a) 3.5 km   (b) 0.025 km   (c) 0.75 km
          (d) 0.9 km   (e) 0.4 km

Page 23 - Exercise 7

Group 25: (a) 5 008 m   (b) 307 cm   (c) 45 mm
          (d) 1 011 mm   (e) 555 cm
Group 26: (a) 642 cm   (b) 5.5 m   (c) 6.25 km
          (d) 3.263 m   (e) 26 cm
Group 27: (a) 7 230.465 m   (b) 15 501 m   (c) 6.666 m
          (d) 3 216 m   (e) 2 513.847 m
Group 28: (a) 5 021.205 m   (b) 703.01 m
          (c) 1 111.111 m   (d) 2.250 m   (e) 2 656.5 m

Page 24 - Exercise 8

Group 29: (a) 6 217.423 m   (b) 14 526 m   (c) 11.748 m
          (d) 60.6 m   (e) 2 513.585 m
Group 30: (a) 4 444 m   (b) 5.834 m   (c) 550.555 m
          (d) 15 212.532 m   (e) 5 432.005 m

ANSWERS TO PRACTICE EXAMPLES (continued)

Page 25 - Exercise 9

Group 31: (a) 160 mm   (b) 26 cm   (c) 920 dm
(d) 350 cm   (e) 3 200 mm

Group 32: (a) 40 m   (b) 155 dam   (c) 310 hm
(d) 375 m   (e) 9 000 m

Group 33: (a) 2.5 cm   (b) 4 dm   (c) 1.1 m   (d) 6 m
(e) 7.5 m

Group 34: (a) 0.75 dam   (b) 1.7 hm   (c) 0.75 km
(d) 1.12 hm   (e) 3.255 km

Page 28 - Exercise 10

Group 35: (a) 500 mm² (b) 6 500 cm²   (c) 2 000 dm²
(d) 46 000 cm²   (e) 16 000 000 mm²

Group 36: (a) 1.5 cm²   (b) 6.98 dm²   (c) 0.05 m²
(d) 5.5 m²   (e) 0.523 m²

Group 37: (a) 750 000 m²   (b) 3 000 000 m²
(c) 5 500 000 m²   (d) 1 300 000 m²
(e) 1 000 000 m²

Group 38: (a) 0.175 km²   (b) 0.25 km²   (c) 0.125 km²
(d) 0.062 5 km²   (e) 0.031 25 km²

Page 29 - Exercise 11

Group 39: (a) 1 100 mm²   (b) 175 cm²   (c) 3 500 dm²
(d) 80 000 cm2   (e) 750 000 mm²

Group 40: (a) 3.72 cm²   (b) 0.88 dm²   (c) 1.587 5 m²
(d) 0.84 m²   (e) 0.6 m²

Group 41: (a) 4 250 000 m²   (b) 5 375 000 m²
(c) 625 000 m²   (d) 1 750 000 m²
(e) 3 375 000 m²

Group 42: (a) 2.25 km²   (b) 1.3 km²   (c) 0.1 km²
(d) 0.037 km²   (e) 0.05 km²

ANSWERS TO PRACTICE EXAMPLES (continued)

Page 32 - Exercise 12

Group 43: (a) 4 000 mm³   (b) 17 000 cm³
          (c) 12 500 dm³   (d) 18 000 000 cm³
          (e) 7 000 000 000 mm³
Group 44: (a) 40 000 mm³   (b) 600 cm³   (c) 50 000 dm³
          (d) 5 500 000 cm³   (e) 1 375 000 000 mm³
Group 45: (a) 2 cm³   (b) 0.65 dm³   (c) 2.52 m³
          (d) 7 m³   (e) 0.006 5 m³
Group 46: (a) 3.25 cm³   (b) 0.0375 dm³   (c) 8 m³
          (d) 0.035 m³   (e) 0.007 m³

Page 34 - Exercise 13

Group 47: (a) 200 l   (b) 250 dal   (c) 64 hl
          (d) 3 700 l   (e) 42 000 l
Group 48: (a) 1.8 cl   (b) 5 dl   (c) 0.7 l   (d) 6 l
          (e) 5.875 l
Group 49: (a) 50 ml   (b) 1.20 cl   (c) 150 dl
          (d) 4 800 cl   (e) 6 600 ml
Group 50: (a) 11.2 dal   (b) 0.277 hl   (g) 4.5 kl
          (d) 2.5 hl   (e) 0.086 kl

Page 35 - Exercise 14

Group 51: (a) 200 ml   (b) 50 cl   (c) 7.5 dl
          (d) 1 500 cl   (e) 25 000 ml
Group 52: (a) 0.9 cl   (b) 7 dl   (c) 2.13 l   (d) 0.75 l
          (e) 2.5 l
Group 53: (a) 70 l   (b) 75 dal   (c) 300 hl   (d) 1 525 l
          (e) 50 000 l
Group 54: (a) 0.7 dal   (b) 5.6 hl   (c) 2.45 kl
          (d) 0.45 hl   (e) 4.95 kl

Page 39 - Exercise 15

Group 55: (a) 50 mg   (b) 230 cg   (c) 120 dg
          (d) 6 000 cg   (e) 6.25 mg
Group 56: (a) 4 cg   (b) 1.5 dg   (c) 0.7 g   (d) 1.5 g
          (e) 2 g
Group 57: (a) 87.5 dag   (b) 360 hg   (c) 1 500 g
          (d) 3 750 g   (e) 4 000 g
Group 58: (a) 12.5 hg   (b) 2.525 kg   (c) 25 hg
          (d) 3.2 kg   (e) 22.25 t

ANSWERS TO PRACTICE EXAMPLES (continued)

Group 59: (a) 52.5 mg (b) 250 cg (c) 80 dg
(d) 1 700 cg (e) 6 500 mg
Group 60: (a) 0.4 cg (b) 4.5 dg (c) 0.725 g
(d) 15 g (e) 0.085 g
Group 61: (a) 350 dag (b) 250 hg (c) 1 225 g
(d) 9 000 g (e) 25 000 kg
Group 62: (a) 2.537 5 hg (b) 2.5 kg (c) 5.55 hg
(d) 3.25 kg (e) 6.5 t

Page 42 - Exercise 17

Group 63: (a) 5 cm^3 (b) 16 cm^3 (c) 35 cm^3
(d) 49 cm^3 (e) 7.1 cm^3
Group 64: (a) 80 cm^3 (b) 190 cm^3 (d) 40 cm^3
(d) 7.5 cm^3 (e) 125 cm^3
Group 65: (a) 400 cm^3 (b) 2 200 cm^3 = 2.2 dm^3
(c) 8 000 cm^3 = 8 dm^3 (d) 750 cm^3
(e) 4 225 cm^3 = 4.225 dm^3
Group 66: (a) 2 dm^3 (b) 75 dm^3 (c) 40 dm^3
(d) 2.8 dm^3 (e) 12.75 dm^3

Page 43 - Exercise 18

Group 67: (a) 4 ml (b) 19 ml (c) 40 ml (d) 6.5 ml
(e) 37.55 ml
Group 68: (a) 0.025 ml (b) 0.015 ml (c) 0.004 ml
(d) 0.002 5 ml (e) 0.007 75 ml
Group 69: (a) 4 l (b) 25 l (c) 46 l (d) 5.75 l
(e) 20.375 l
Group 70: (a) 5 000 l - 5 kl (b) 7 kl (c) 19 kl
(d) 875 l = 0.875 kl (3) 13.5 kl

Page 44 - Exercise 19

Group 71: (a) 5 kg (b) 22 kg (c) 7.75 kg (d) 15 kg
(e) 3.5 kg
Group 72: (a) 7 kg (b) 11 kg (c) 7.8 g (d) 15 g
(e) 56 g
Group 73: (a) 3 kg (b) 85 kg (c) 11 kg (d) 5.25 kg
(e) 14.75 kg
Group 74: (a) 9 g (b) 45 g (c) 10.5 g (d) 22.5 g
(e) 30 g

ANSWERS TO PRACTICE EXAMPLES (continued)

Page 45 - Exercise 20

Group 75: (a) 7 l = 7 dm³   (b) 35 l = 35 dm³
            (c) 10 l = 10 dm³   (d) 8.25 l
            (e) 0.37 l = 370 ml = 370 cm³
Group 76: (a) 6 ml   (b) 15 ml   (c) 72 ml   (d) 0.35 ml
            (e) 30.25 ml
Group 77: (a) 5 dm³   (b) 12 dm³   (c) 90 dm³
            (d) 2.8 dm³   (e) 25.67 dm³
Group 78: (a) 6 cm³   (b) 35 cm³   (c) 48 cm³
            (d) 3.2 cm³   (e) 62.5 cm³

Notes

Notes

Notes

Notes

Notes

Notes

Notes